REDUCING BR
FATIGUE FAILURES IN STEEL STRUCTURES

Peter Maranian, S.E.

SPONSORED BY
Technical Council on Forensic Engineering (TCFE)
of the American Society of Civil Engineers

ASCE AMERICAN SOCIETY OF CIVIL ENGINEERS
1801 ALEXANDER BELL DRIVE
RESTON, VIRGINIA 20191–4400

Cataloging-in-Publication Data on file with the Library of Congress.

American Society of Civil Engineers
1801 Alexander Bell Drive
Reston, Virginia, 20191-4400

www.pubs.asce.org

ISBN 978-0-7844-1067-7
Manufactured in the United States of America.

Foreword

After the Northridge (California) Earthquake of 1994, many engineers and researchers took a careful look at failures caused by that major seismic event. In 1995, the Kobe (Japan) Earthquake revealed similar problems, principally in steel moment-frame structures. Among the experienced and talented engineers looking at these problems were many members of ASCE.

Within ASCE, the Technical Council on Forensic Engineering (TCFE) is comprised of members from across the country who came together years earlier around common elements of their practices in forensic engineering. Principal among the TCFE tenets has always been the desire to help prevent future failures by reporting as much as could be learned about previous failures. It is not enough to learn from our own mistakes, we progress as a profession when we also learn from the mistakes of others.

Within TCFE, the Committee on Practices to Reduce Failures (CPRF) includes several structural engineers who practice in California, and who engaged in the post-earthquake examination and review of damaged steel structures. Henry Huang and Peter Maranian, two of the engineers intimately involved with the analysis of the damaged steel structures, both came to realize how much they had learned through their investigations. From this shared experience came the idea for a book that would record their thoughts, point others to relevant references, and provide a first step in an engineer's search for a deeper understanding of the problems of fatigue and brittle fracture. During Henry's time as chair of the CPRF, he encouraged Peter to put their newfound knowledge into a publication for ASCE.

This book represents the culmination of Peter's work to date. He drew heavily from his and Henry's own experiences, but as his acknowledgments reflect, he also benefited from the tangible contributions of many colleagues, and from the support of CPRF and TCFE. Many fellow committee members have reviewed this document at various stages during its development. However, the work is principally Peter Maranian's, and represents a milestone in his journey. We trust that it will provide the reader with an excellent primer on the subject of brittle and fatigue failures in steel structures. We are honored to have played a small part in bringing this book to press.

Leonard J. Morse-Fortier, PhD, SECB
Chair, TCFE

Preface

Structural steel has proven to be an excellent and versatile building material that has enabled small to very large structures to be constructed for the great benefit of society. However, there have been repeated failures associated with fracture and/or fatigue mechanisms.

Failures of steel moment frame connections in the 1994 Northridge Earthquake in California, USA and the 1995 Kobe, Japan are well documented. The 1994 Northridge Earthquake led to a significant amount of rethinking, testing and research by the Structural Engineers Association of California (SEAOC), the Applied Technology Council (ATC) and California Universities for Research in Earthquake Engineering (CUREe), called the SAC Joint Venture for the Federal Emergency Management Agency (FEMA) [which resulted in the publication of several documents including FEMA 350 (2000), 351 (2000), 352 (2000), 353 (2000) and other documents] and several other independent organizations. As a result, many engineers have not only questioned seismic designs for steel moment frames, but also other steel seismic resisting systems (e.g. braced frames and eccentric braced frames) along with buildings subjected to high winds. Although at the time of writing this publication, the causes of the Mississippi River Bridge collapse in August 2007 are not known, it highlighted concerns for the life of bridge structures particularly with regard to fatigue and corrosion.

The problems of brittle and fatigue failures in steel along with corrosion have gone on for the most part of the last century. Factors and issues affecting failures have been, in many cases, well researched and documented over several decades. However, these factors and issues have not always been translated into state of the art design practices. Experiences with failures found in one industry, which may have resulted in changes in that industry, have not always affected changes in related industries.

The intention of this book is to describe the characteristics of steel and associated fabrication processes identifying many of the potential problems that can lead to fracture. It is hoped that the publication will help give engineers a better understanding of steel, its limitations and applications, in order to reduce brittle and fatigue failures.

Chapter 1 discusses examples of failures, most of which took place in the last century. Chapter 2 gives a brief introduction to fracture mechanics including the concepts behind the need for design considerations to minimize stress and strain concentrations, quality control and assurance, and adequate material properties. Chapter 3 discusses steel as a material including the processes, chemistry, and mechanical properties. Chapter 4 discusses fabrication and connections including the effects of fabrication procedures, welding, bolting, and riveting on the finished product. Chapter 5 gives a discussion and recommendations regarding addressing brittle and fatigue failures in steel buildings. Also included in Chapter 5 are discussion on the issues and recommendations for current practice.

It should be noted that this document primarily addresses brittle and fatigue type of failures and issues associated with corrosion. This document does not address failures of steel structures due to instability and erection procedures. Further reading of the topics in this document is encouraged. Recommended reading is provided at the end of each section and references at the end of the document.

Peter Maranian, S.E.

Acknowledgments

This publication was carried out for the Committee on Practices to Reduce Failures (CPRF) which is one of the committees of the American Society of Civil Engineers' Technical Council for Forensic Engineering (TCFE). It has taken several years of effort to compile, and could not have been carried out without the significant contributions, consultation, advice and support of many. My very sincere thanks to all involved.

In particular, I wish to acknowledge the major contribution made by Andrew Metzger who helped write most of the very important discussion on Fracture Mechanics given in Chapter 2. Andrew Metzger's fine contribution and knowledge on this key subject is so very much appreciated.

I also wish to acknowledge Eric Stovner, who made an important contribution to the case histories given in Chapter 1, describing the Mianus River Bridge failure. Eric, when chair of CPRF, also played a key role in progressing the document including carrying out an extensive review of the manuscript.

Furthermore, I can not thank enough both Leonard Morse-Fortier, chair of TCFE from 2008 to 2009, along with Susan Sheldon Smith for their great efforts in transforming the final original manuscript and the collection of figures into the book you see here. Leonard's great abilities, leadership and devotion to the task along with Susan's significant talents all were applied in the final compilation and editing of the document. Their contributions were immeasurable.

The document was the brain child of Henry Huang with whom I have worked for many years on various committees. Henry, at the time, was chair of CPRF. He and I worked together investigating problems with steel structures damaged by the 1994 Northridge (California) Earthquake. Henry saw the important need to have a document on structural steel that would convey the concerns that we shared after those experiences. His great vision and understanding of the issues effecting the profession were an inspiration to me.

I had the benefit of some excellent and constructive review comments from several members of CPRF. These included Rubin Zallen, whose thorough and critically constructive review is sincerely appreciated. Michael Lester, John Gross, and Ron Schneider also provided important reference material. I appreciate the thorough review comments from members of the TCFE's executive committee, including those by Norb Delatte, Anthony Dolhon and David Peraza. Michael Lester, who took over the chair of CPRF after Eric Stovner, also greatly assisted in the administrative aspects of developing the manuscript.

Along with Henry Huang, there were several others who shared many of their concerns regarding steel issues and gave invaluable support. The late Dr. Warner Simon was a major inspiration to me providing much insight on the problems. Although well into his eighties, I was privileged to work with him during his final years, throughout which he dedicated himself to the pursuit of investigating past failures and resolving welding issues. I was also greatly privileged to have the opportunity to assist Dr. James Anderson at the University of California, Los Angeles, on several steel moment frame connection tests carried out during the 1990's from which I gained invaluable knowledge. Dr. Praful Patel provided excellent knowledge on issues associated with heat affected zones and assisted with reviewing the document. I also had the significant benefit of working with Daniel Luna, a welding consultant, who has become a valuable resource on welding procedures and processes. I was also very fortunate to have Robert Lyons as a colleague, both at the company I work with, and in committees. His help included taking on the burden of articulating many of the issues through committees. Ashwani Dhalwala, particularly during recent years, has been a great resource, providing much insight, knowledge along and encouragement on addressing the issues of brittle failures. Others who have helped by providing invaluable documents include James Partridge, Raphael Franco, Dr. Patxi Uriz, Saif Hussain and Don Strand. Throughout all of these efforts, Dr. Gregg Brandow, President at Brandow & Johnston Inc, gave me constant support, encouragement and also helped provide some important references.

There are several people whose hard work much helped to produce the document and are much appreciated. They include Agnes Gonzalez who carried out the typing dealing with countless drafts, Sema Akyurek who carried out most of the sketches and illustrations along with Marco Ramirez, Gabriel Lopez who assisted in organizing the sketches, Dr. Navid Nastar who provided much assistance in reviewing and checking the document and Bhahti Dhalwala who carried out an editorial review of the document.

Finally, I wish to thank my wife, Srpuhi, for her patience and understanding in tolerating endless weekends during which I worked on the document.

Peter Maranian, S.E.

List of Figures

Contents

CHAPTER 1

EXAMPLES OF MAJOR HISTORICAL EVENTS

Brittle steel failures have repeatedly occurred during the past century. A brief outline of some of the major events is as follows:

1.1 EVENTS WITH NON-WELDED CONNECTIONS

1.1.1 Riveted Standpipe, Long Island, New York

A failure occurred during the hydrostatic testing of a riveted standpipe 76m (250 ft.) high in Long Island, New York in 1886 [Parker (1957)]. The diameter of the standpipe was 4.88m (16 ft.) up to a height of 18m (59 feet) and decreased conically to a diameter of 2.44m (8 ft.) at a height of 25.6m (84 ft.), Plates of 1.52m (5 ft.), by 2.14m (7 ft.) and 1.52m (5 ft.), by 2.75m (9 ft.) were used with thicknesses varying from 25 mm (1 inch) at the bottom to 6 mm (¼ inch) at the top. The plates were connected by rivets and the standpipe was stabilized by guy wires. A vertical crack approximately 6.1m (20 ft.) long near the bottom of the standpipe occurred during a hydrostatic test when the water level had reached a height of 69.2m (227 ft.) and resulted in immediate and total collapse of the structure. The report, in the October 23, 1886 edition of Engineering News, described utter destruction with a clean cut just below the cone "likened to nothing better in effect than the sudden smashing of a high glass cylinder." The report further describes "brittle material concentrated in the portion of the tower exposed."

1.1.2 Titanic, North Atlantic Ocean

Following the recovery of the wreckage of the Titanic in 1985 in the northern Atlantic Ocean and the recovery of a piece of the hull in 1991, a panel of investigators concluded, pending further information, that the sinking of the Titanic in 1912 was due to the fracture of wrought iron rivets, in low temperature water, which led to the parting of the hull plates resulting in flooding of the ship. Examination of rivets found that they contained as much as 9.3% silicate slag in the form of stringers more than 200μm in length greater than the 2% to 3% slag normally permitted in wrought iron rivets at that time. It was also found that the wrought iron had relatively high carbon and sulphur content (0.21% and 0.065% respectively) and low manganese content (0.48%). These resulted in detrimental effects on impact toughness. [McEvily (2001), Gartzke, et al (1992), McCarty and Foecke (2007)]

1.1.3 Molasses Tank, Boston, Massachusetts

A molasses tank, using riveted connections, fractured in January 1919 in Boston, Massachusetts with twelve people drowning in molasses and forty other people injured [Parker (1957)]. Several houses were damaged, many horses drowned and a portion of the Boston Elevated Railway Structure was destroyed. Stresses of between 276 MPa (40,000 p.s.i.) to 345 MPa (50,000 p.s.i.) were calculated at the riveted joints. The tensile strength of the steel was 379 Mpa (55,000 p.s.i.). Herringbone patterns in the failure surfaces of the steel from the tank indicated fracture resulting from notches in the plates.

1.1.4 Brittle Steels in Belgium

Brittle steels were encountered in pre-World War II steel fabrication in Belgium when in 1934 a girder cracked, along its complete length, in the fabricator's yard, prior to being subjected to load (see Figure 1.1a). The day before the fracture, the ends had been cut at a skew using a cutting torch. At about noon, a violent noise occurred giving the impression of an explosion;

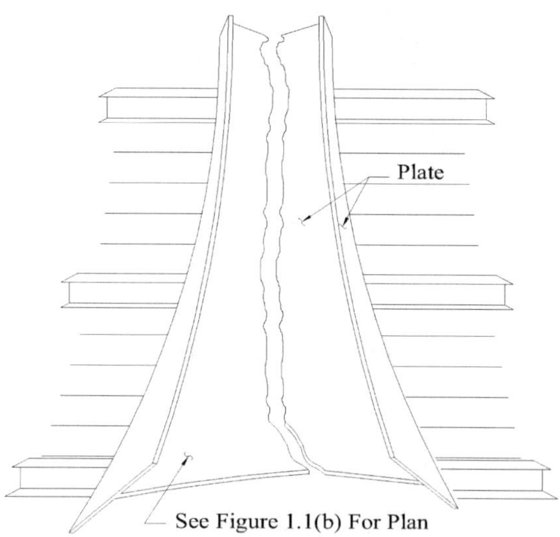

3D View

Steel Plate Girder in Belgium, Cracked 1933

Figure 1.1a

the result of the girder fracturing along its entire length in a very rough and jagged manner. The flanges deflected inwards as shown in Figure 1.1b. Campus (1954), who had designed the girders for the University of Liege, suggested that high residual stresses, as a result of restraint of the welding of the flanges to the web, was the main cause. However, Lancaster (1992) states that the steel, made by the Bessemer Process used nitrogen that led to the strain-age embrittlement. Both theories may be correct. Campus also reported other fractured girders and built up sections. Figure 1.1c shows a cracked wide flange "grey-beam" that fractured due to small cuts being made at the ends of the flanges. The figures and illustrations were drawn from review of photographs that can be found in Campus (1954) and Parker (1957).

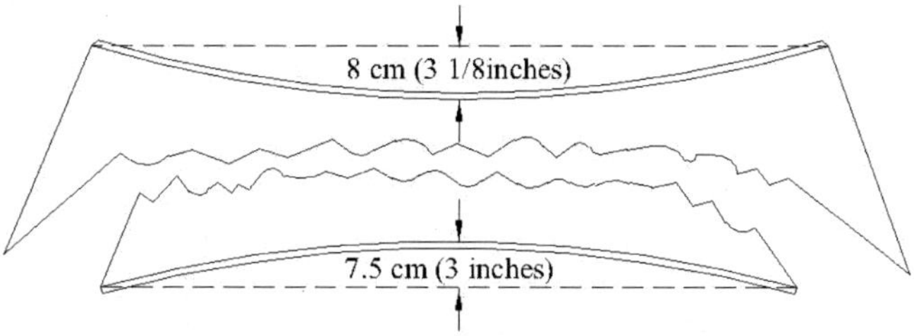

Plan

Steel Plate Girder in Belgium, Cracked 1933

Figure 1.1b

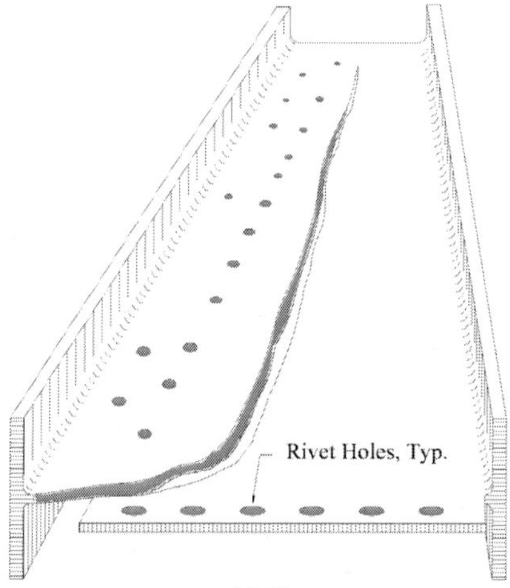

3D View

Wide Flange Grey Beam, Belgium 1930's

Figure 1.1c

1.1.5 Point Pleasant Suspension Bridge, West Virginia

The Point Pleasant Suspension Bridge collapse on December 15, 1967 in West Virginia, 41 years after construction, was apparently due to stress corrosion and/or corrosion fatigue of an eyebar which was an essential part of the suspension member (See Figures 1.2a and 1.2b). The lower limb of the eyebar

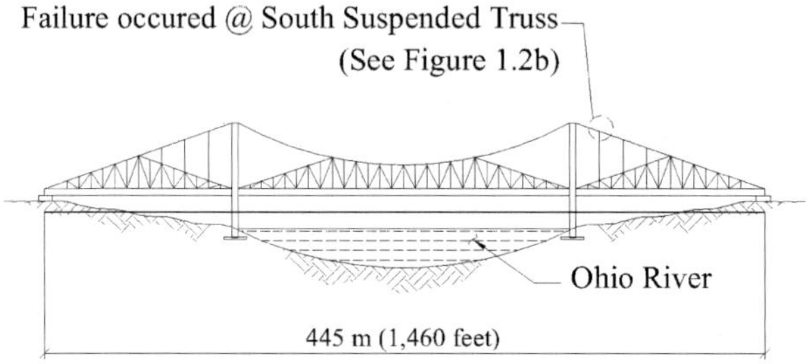

Elevation Looking South

Point Pleasant Bridge Failure, West Virginia, 1967
[From "Why Buildings Fall Down" by Mathys Levy and Mario Salvadori, Copyright 1992 Levi and Salvadori. Used with permission of W.W.Norton & Company Inc.]

Figure 1.2a

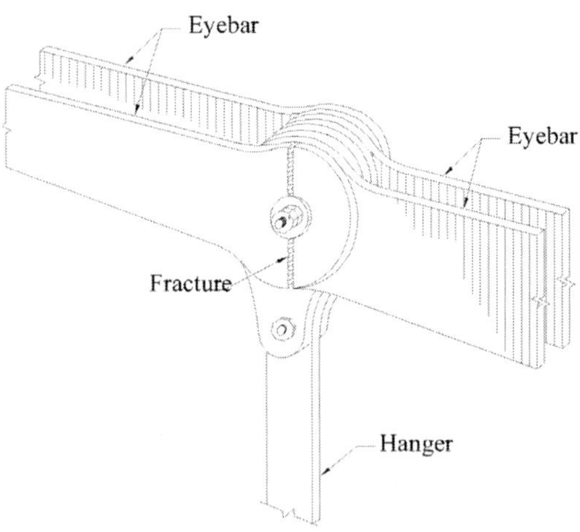

3D View of Cracked Eyebar

Point Pleasant Bridge Failure, West Virginia, 1967
[From McEvily (2001), reproduced with permission of John Wiley and Sons, Inc.]

Figure 1.2b

fractured first, followed by fracture of the upper limb of the eye. Forty-six people were killed in the accident, which occurred around 5:00 p.m. The eyebars, approximately 9.2m (30 feet) long and arranged in pairs, were not redundant. That is, failure of one bar would cause failure of the other bar. Extremely high stress concentrations occurred in the eyebar caused by friction at the pin with high stress concentrations at corrosions pits. Corrosion was also accentuated by exposure to air pollutants such as sulphur dioxide. The temperature was approximately -1°C (30°F). [McEvily (2001), Levy and Salvadori (1987), Sih (1989)]

1.1.6 Ingram Barge, Long Island, New York

A brittle fracture of the 178m (584 ft.) long Ingram Barge occurred in 1972 in Port Jefferson Harbor, Long Island, New York (see Figure 1.3a). The three-dimensional sketch shown in Figure 1.3a was drawn from review of photographs in Barsom and Rolphe (1999) and Masabuchi (1980). The material had been tested and demonstrated good notch toughness. However, failure was apparently due to high stresses caused by improper ballasting [Masubuchi (1980)]. Barsom and Rolphe (1999) state that unusually high stresses, 2.5 times design loading, occurred when the barge was turning in the harbor in calm waters with air temperatures at -14°C (7°F).

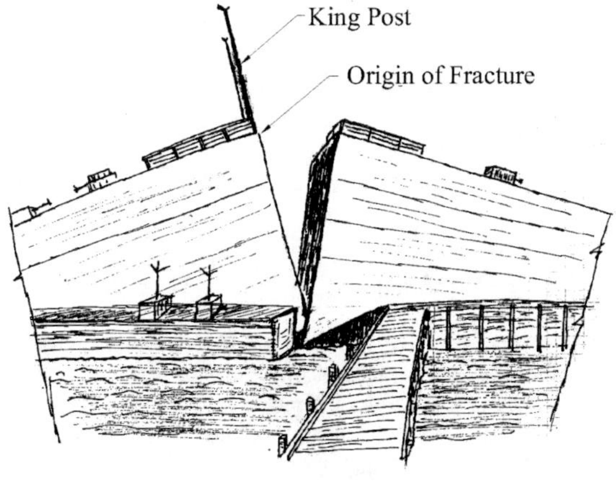

Ingram Barge Fracture, Long Island, New York, 1972
(Reproduced with permission from Fracture and Fatigue Control in Structures, by Barsom and Rolphe, 3rd Edition, copyright, ASTM International)

Figure 1.3a

The initiation of the fracture, according to Barsom and Rolphe, occurred at a doubler plate welded to the deck plate with the king post welded to the doubler plate (see Longitudinal Section, Figure 1.3b).

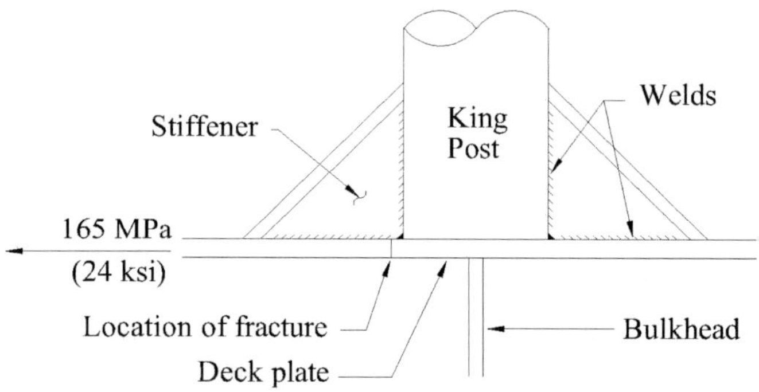

Longitudinal Section

Ingram Barge Fracture, Long Island, New York, 1972
(Reproduced with permission from Fracture and Fatigue Control in Structures, by Barsom and Rolphe, 3rd Edition, copyright, ASTM International)

Figure 1.3b

The conclusion was that the improper ballasting caused stresses exceeding the capacity of the material at the highly restrained region. The longitudinal stresses of 165 MPa (24 k.s.i.) were not excessive. This event demonstrated the potential for brittle failure to occur due to high restraint at moderate stresses.

1.1.7 Citicorp Plaza, New York

The substitution of bolted connections for welded connections, due to cost considerations, was accepted during the shop drawing phase of the project (circa 1978). Although bolted connections were permitted, subsequent to completion of the construction, the Engineer of Record carried out another analysis of the building. This time he included quartering winds (i.e. at 45° to the axis of the building), to check if earlier decisions, made by his associates, were justifiable (see Figure 1.4). His analysis surprisingly showed that the bolted connections were as much as 160 percent overstressed. Had the joints been welded as originally planned, there would not have been a problem. Much to his credit, the Engineer of Record made the very bold and politically difficult steps to initiate retrofit of the connections by welding [New Yorker (1995)].

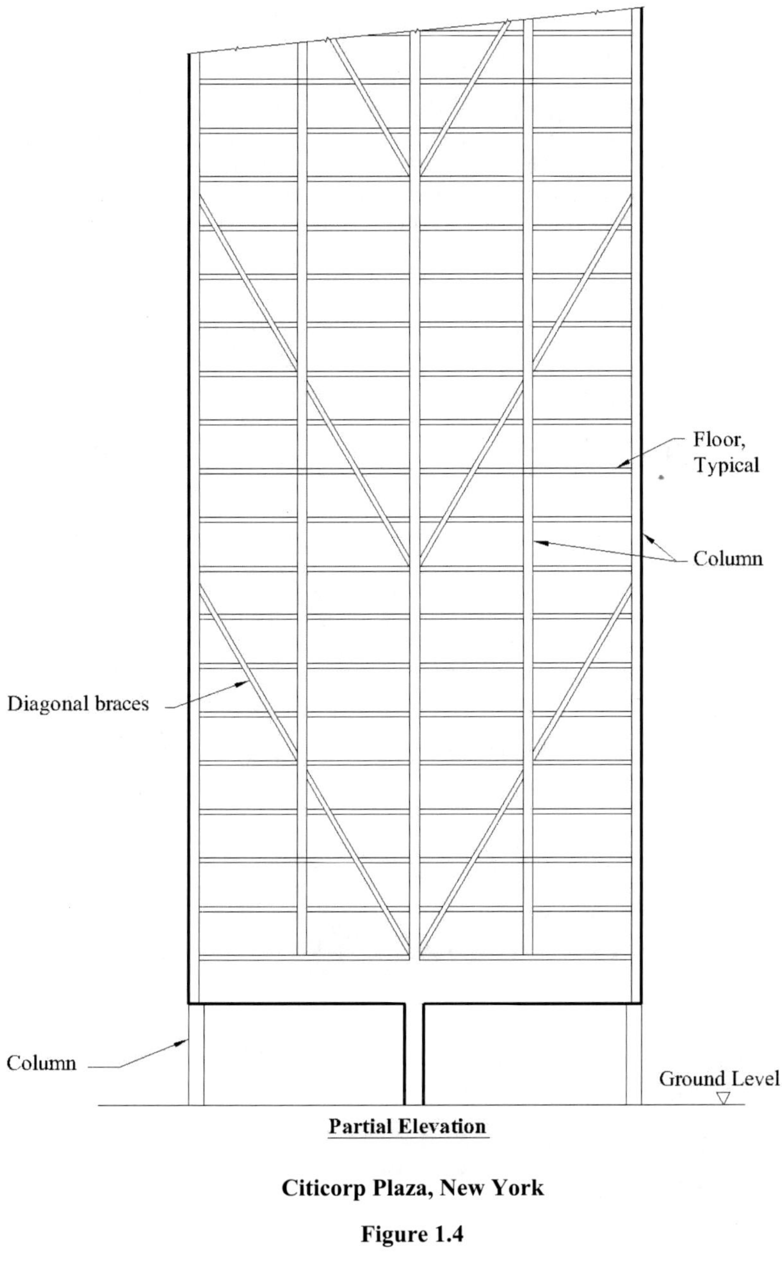

Partial Elevation

Citicorp Plaza, New York

Figure 1.4

1.1.8 Mianus River Bridge, Connecticut

The Mianus River Bridge, a vehicular multi-span bridge of cantilever construction, opened in Greenwich, Connecticut in 1958 and failed on June 28, 1983 (see Figure 1.5). The bridge is 809.5m (2656 feet) long approximately 21.4m (70 feet) above marshland and tidal flats. A 30.5m (100 foot) long span collapsed killing three people and injuring another three. The suspended span consisted of a 19 cm (7½ inch) concrete slab on several cross beams, four stringers and two 2.75m (9 foot) deep plate girders. At each side of the

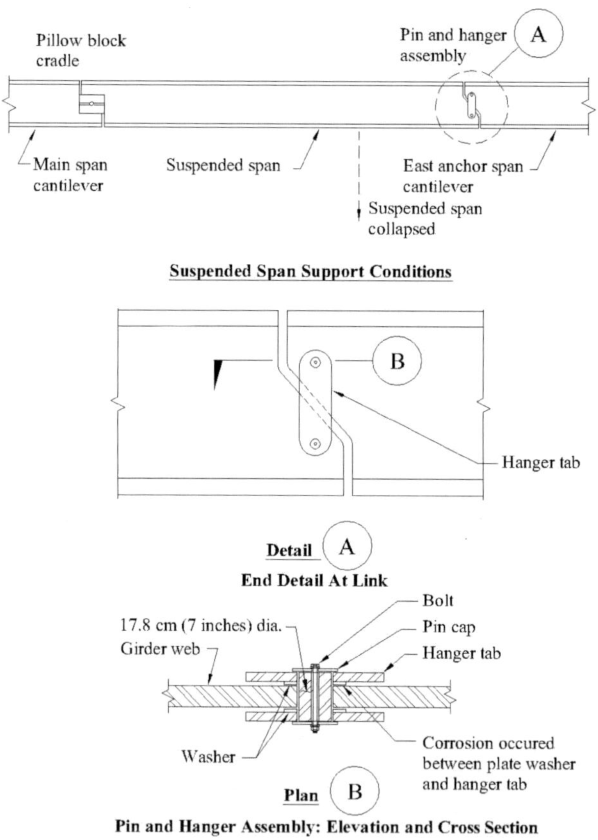

Mianus Bridge Failure, Connecticut, 1983

[From "Why Buildings Fall Down" by Mathys Levy and Mario Salvadori, Copyright 1992 Levi and Salvadori. Used with permission of W.W.Norton & Company Inc.]

Figure 1.5

collapsed suspended span was a plate girder connected at its ends to the cantilevered girders. The cantilevered supports were skewed in plan. The connections at the collapsed end consisted of plate hanger assemblies with plates each side of the girder webs having one 17.8m (7-inch) diameter pin through the suspended girder and one 17.8 cm (7-inch) diameter pin through the cantilevered girder (see details A & B, Figure 1.5). Ten years prior to the collapse, roadway drains were covered, and the hanger assemblies were subjected to excessive water infiltration. Corrosion occurred in the plate washers between the washer and hanger tab (and not readily observable during bridge inspections), causing extensive corrosion at the interface between the hanger plate and the pin. This resulted in a reduction in the thickness of the hanger plate. Furthermore, the build up of corrosion caused expansive forces to occur that popped a plate washer and nut off of a lower pin. Loads were doubled to the remaining effective hanger plate, the corner of the suspended span dropped slightly, and loads were redistributed eccentrically to the upper pin. The repeated pounding of traffic caused a fatigue crack in the upper pin that eventually propagated leading to collapse of the span [Levy and Salvadori (1992)].

1.1.9 1994 Northridge Earthquake, California

The January 17, 1994 Northridge Earthquake in California was mostly known for failures of steel moment frame connections [see 1.2.11]. However, there were also examples of bolt failures. Figure 1.6 shows failure of all bolts at a shear plate connecting a tilt up wall to a steel column in a three-story building. The ASTM A325 bolts sheared in a brittle manner without much display of ductility.

There were incidences involving damage to braced frames including local buckling of steel tube braces, tearing of steel at corners of steel tubes and complete rupture of braces. These are shown in Figures 1.7a, 1.7b and 1.7c. It is interesting to note that similar types of failures occurred in a test on a two-story special concentric braced frame structure using rectangular hollow steel sections carried out at the University of California Berkeley by Uriz and Mahin in 2004 [Uriz and Mahin (2004) and SEAOSC (2005)].

There was also brittle failure of base plates at braced frames observed from the 1994 Northridge Earthquake [Bertero et al (1994)].

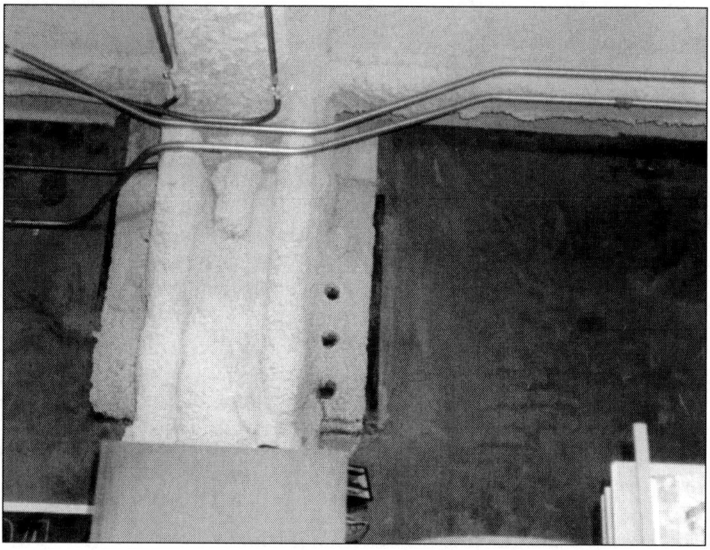

Shear failure of bolts at shear plate at tilt-up wall
Northridge Earthquake, Los Angeles, 1994

Figure 1.6

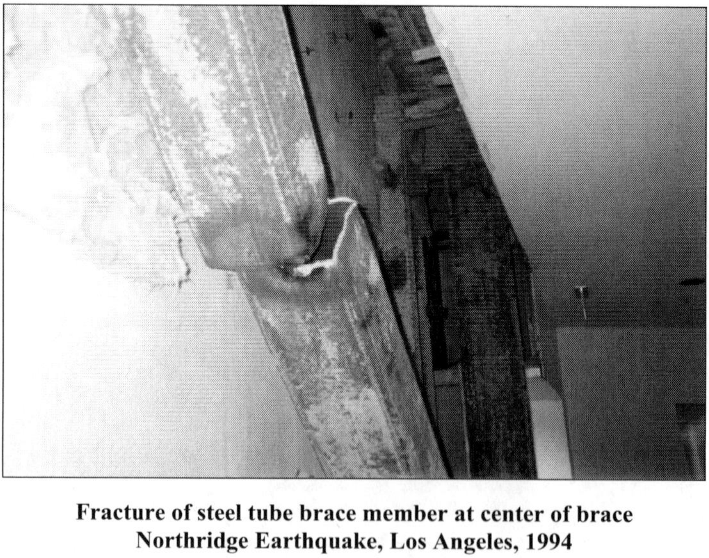

Fracture of steel tube brace member at center of brace
Northridge Earthquake, Los Angeles, 1994

Figure 1.7a

Fracture of steel tube brace member at center of brace
Northridge Earthquake, Los Angeles, 1994

Figure 1.7b

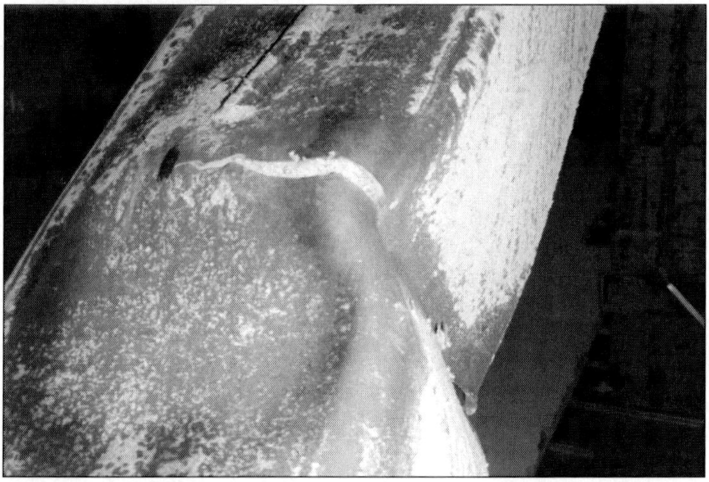

Local buckling of steel tube brace member at center of brace
Northridge Earthquake, Los Angeles, 1994

Figure 1.7c

1.1.10 1995 Hyogoken-Nanbu (Kobe) Earthquake, Japan

There was widespread damage to braced frames particularly in the pre-1970 buildings which had slender braces from the January 17, 1995 Hyogoken-Nanbu (Kobe) earthquake [A1J (1995)]. The damage included buckling of braces, rupture of flat plate and wide flange bracing members, large deformation and tearing of gusset plates (see Figure 1.8a). There were also several cases of fracture of the fabricated box columns within 20.3 cm (8 inches) of welds (see Figures 1.8b and 1.8c). Some damage resulted in partial collapse of buildings. The figures were drawn from photographs that can be found in AIJ (1995).

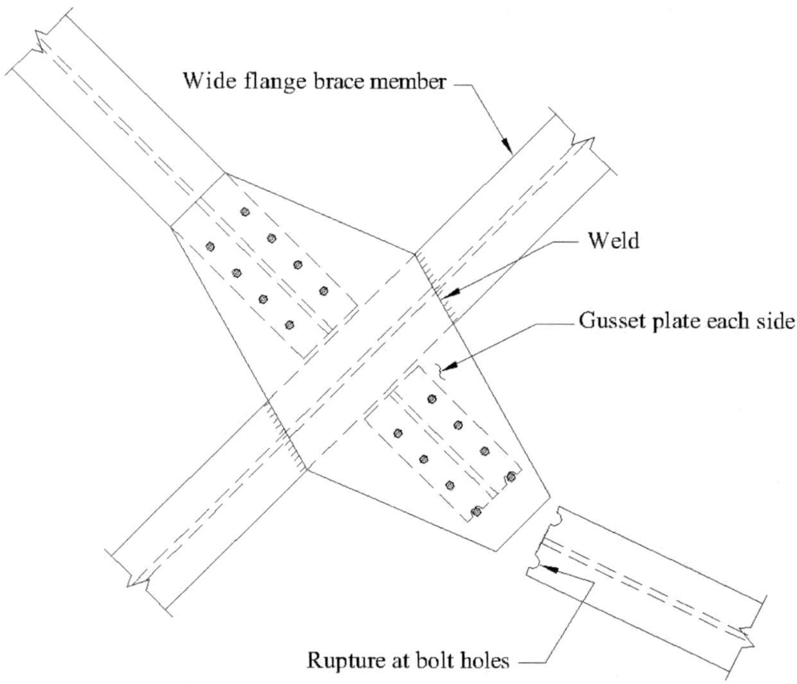

Elevation

Rupture of Cross Bracing Member at Gusset Plate
Hyogoken-Nanbu (Kobe) Earthquake, Japan, 1995

Figure 1.8a

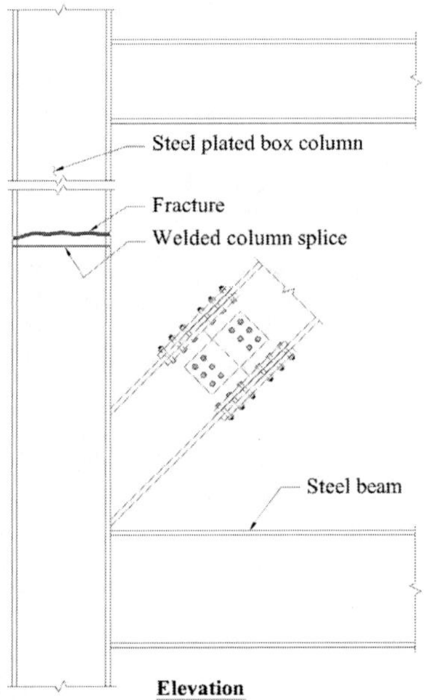

Fracture Near Welded Column Splice of a Braced Frame
Hyogoken-Nanbu (Kobe) Earthquake, Japan, 1995

Figure 1.8b

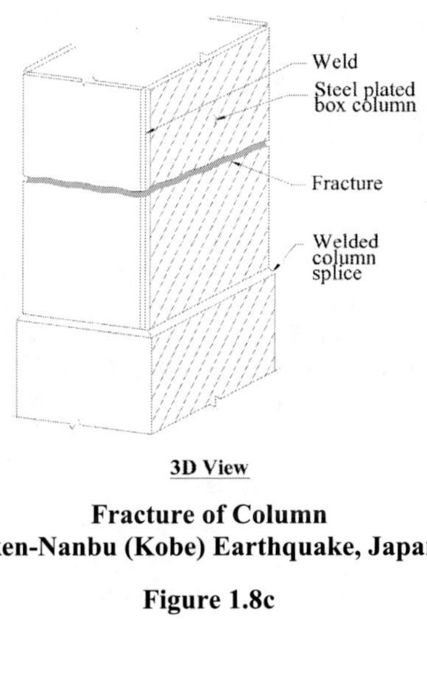

Fracture of Column
Hyogoken-Nanbu (Kobe) Earthquake, Japan, 1995

Figure 1.8c

1.1.11 I-35 Mississippi River Bridge Collapse, Minnesota

The I-35 Mississippi River Bridge collapsed just after 6:00 p.m., during rush hour on August 1, 2007 with the loss of 13 lives and many injured. The bridge was one of two major river crossings [Hansen (2007)].

The bridge, constructed between 1964 and 1967, carried eight lanes and was 581.3m (1907 feet) long, with fourteen spans. The center and longest span was 140m (458 feet) with the adjacent spans 81m (266 feet) supported on 11.6m (38 feet) cantilevers from the approach spans. These spans comprised two steel arched trusses, up to 18.3 (60 feet) deep at the piers. However, the truss configurations were such that there was a lack of redundancy so that failure of any member would lead to collapse of the truss. Welded floor beam trusses connect the two main trusses and extended out to support roadway stringers comprising 68.7 cm (27 inch) deep rolled steel beams, which in turn supported a reinforced concrete deck. The approach spans mostly comprised welded plate girders supporting the deck.

At the time of the disaster, traffic was moving slowly, limited to four out of the eight lanes since the remaining four lanes were closed for resurfacing. Construction vehicles were located on the bridge at the time of the collapse, contributing to significant loads. The construction involved repairs to the concrete deck and expansion joints. The center span suddenly gave way and collapsed into the river and riverbanks.

At the time of writing, the causes of the collapse have not been fully identified. However, initial onsite investigation indicated failure of gusset plates at the top chord and near the quarter span of the main truss early during the event [Holt and Hartmann (2008)]. Analysis by Holt and Hartmann found principal compressive stresses in some of the gusset plates to be as much as 15% overstressed. According to Civil Engineering (Feb. 2008), sixteen gusset plates, eight each side of nodes, were found to be fractured. The remaining gusset plates were found to be intact.

The bridge did have a history of discoveries, which, over its life, accumulated concerns. Significant corrosion of its bearings was cited by the Federal Government in 1990. The bridge suffered from the severe cold weather, which created black ice on the surface, causing several accidents. As a consequence, magnesium chloride was tested in January 1999 and October 1999; potassium acetate was applied using an automated de-icing system to prevent black ice.

In 1998, fatigue cracks were found in several girders near the ends of the approach spans. The web cracked through entirely at one location. Remedial work was carried out including drilling holes at the ends of cracks and reinforcing with bolted plates. Out of plane distortion of the cross girders was occurring which led to cracking. Struts were added to reduce the distortion and holes were drilled to prevent further propagation of cracks. In 2000, loss of sections due to corrosion and fatigue cracks was found in the floor beams at the truss ends and approach span beams. A hinge joint at an approach span was also found to be locked in 2000 [NCE International (2007)]. More fatigue cracks were found in 2004. The National Inventory Database, in 2005, gave the bridge a low rating, saying that the bridge was structurally deficient and only provided minimum tolerable limits if it was to be left in place as is. In a 2006 bridge inspection report [Pribula (2006)], the center and adjacent spans were described as "Fracture Critical". The report by Pribula (2006) described surface rust corrosion and severe section loss at the main truss members. Also, with regard to the main trusses, the report noted numerous poor weld details, and tack welds that had cracked. Pribula, with respect to the floor beam truss members, also described poor weld details, undercut in welds, offsets in top chord splices causing bending of splice plates, severe section loss and pitting at the main truss connections. The report by Pribula recommended replacement of the entire structure or re-decking if replacement was to be significantly delayed. In 2007, recommendations for strengthening 52 fracture critical truss members (typically the top and bottom chord members in tension) were made by a consultant, along with removal and repair of weld defects [NCE International (2007)].

Although the causes of the failure have not yet been identified at the time of writing, fatigue (from heavy trucks, wind and/or construction machinery) and corrosion (associated with severe weather conditions), addressed in this publication, could be participant factors.

1.2 EVENTS WITH WELDED CONNECTIONS

1.2.1 Welded Bridges in Belgium

Several pre-World War II welded steel vierendeel bridges in Belgium collapsed without any live load. Campus (1954) describes several welded bridges built between 1932 and 1938 and reported that the first bridge collapsed in March 1938. The first spectacular failure occurred in March 1938 at the bridge, spanning 74.7m (245 feet), over the Albert Canal in Hasselt, Belgium. Here, after only one year in service, the bottom chord of a vierendeel truss failed at the weld of a cover plate under the load of a tramcar during cold weather [-20°C (-4°F)]) (see Figure 1.9a). The third and fourth verticals also parted. The top chord held for six minutes and eventually the bridge broke into three pieces and fell into the canal. Some fractures passed through welds and others through plates. The chords were made up of plate thicknesses varying from 19mm (¾ inch) to 56mm (2-3/16 inch). Details of other pre-World War II bridge failures were also reported by Boyd (1970) and Parker (1957). The Herenlhais-Oalen Bridge in Belgium, spanning 61m (200 feet), failed on

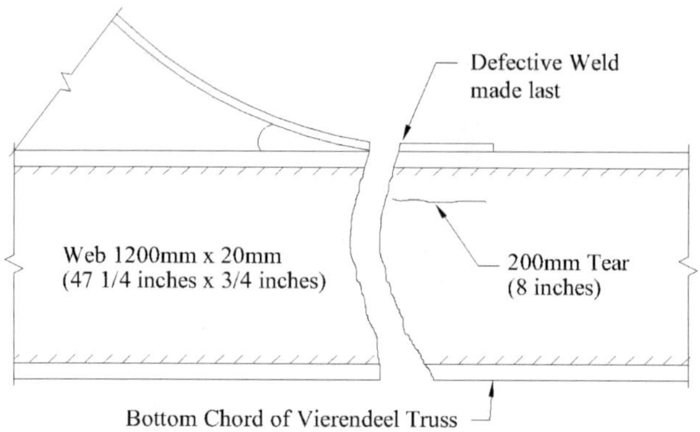

Detail at bottom chord

Hasselt Bridge Failure, Belgium, 1938

Figure 1.9a

January 19, 1940, three years after it was built at a temperature of about -14° C (7°F). Three loud reports were heard but the bridge did not collapse. Five hours later, a locomotive traversed it without causing failure. Subsequent inspections found several cracks all at welded junctions [Parker (1957)]. The 48.8m (160 foot) bridge in Kaulille, built in 1935, failed on January 25, 1940 also at a temperature of -14° C (7°F). Again, the bridge did not collapse but several cracks were found in the lower chord [Parker (1957)]. The failures of the Herenlhais-Oalen Bridge are shown in Figures 1.9b, 1.9c and 1.9d. Figures 1.9e, 1.9f and 1.9g show the failures at the Kaulille Bridge. Subsequent

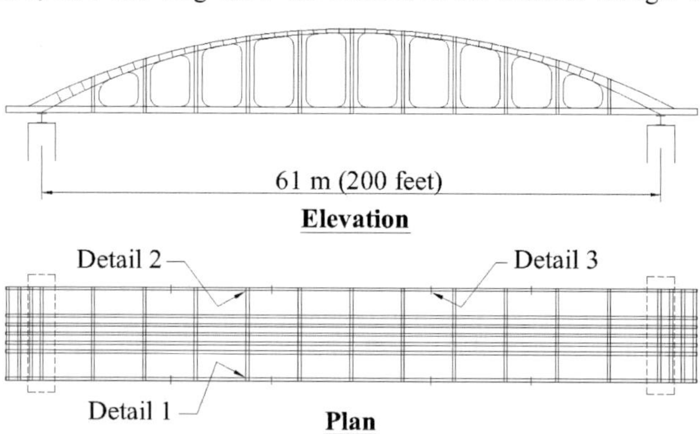

Herenlhais-Oalen Bridge Failure, Belgium, 1940

Figure 1.9b

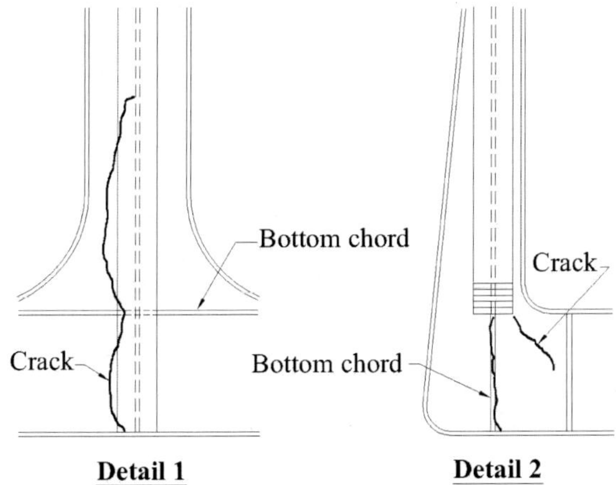

Herenlhais-Oalen Bridge Failure, Belgium, 1940

Figure 1.9c

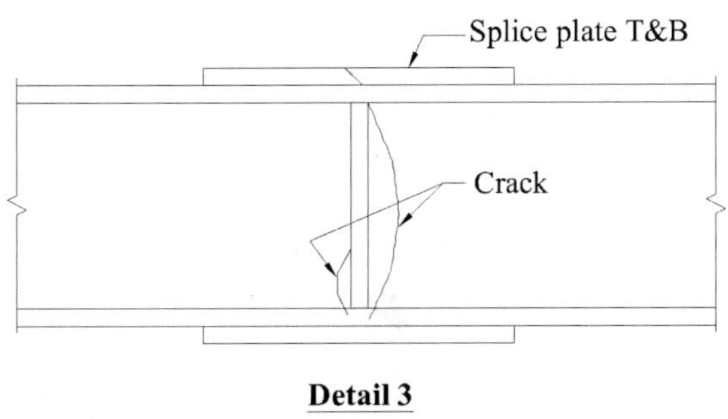

Herenlhais-Oalen Bridge Failure, Belgium, 1940

Figure 1.9d

mechanical tests were carried out on specimens abstracted from the bridges. The steel used for these bridges, made from the Bessemer process, had higher contents of phosphorus and sulphur than normally found in steel, made at that time from the open hearth process (see Chapter 3 for description of processes). Bend tests on plates with longitudinal welds failed when bent through only a small angle [Parker (1957)]. Campus (1954) was of the opinion that several welds had large defects, which were caused by high residual stresses. The figures were drawn from a review of illustrations in Boyd (1970) with the intent to more clearly show the bridge details and failures.

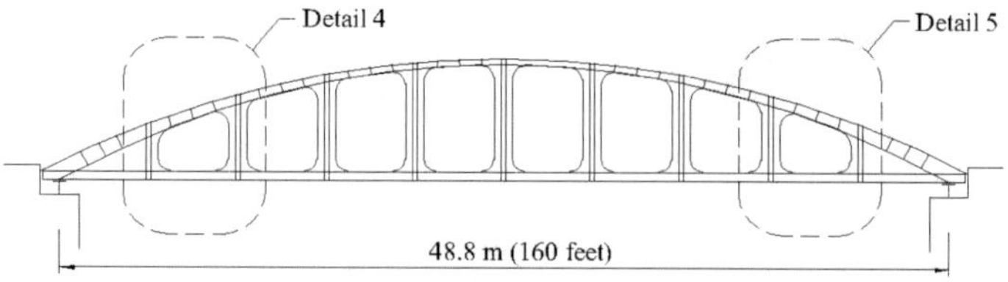

Elevation

Kaulille Bridge Failure, Belgium, 1940

Figure 1.9e

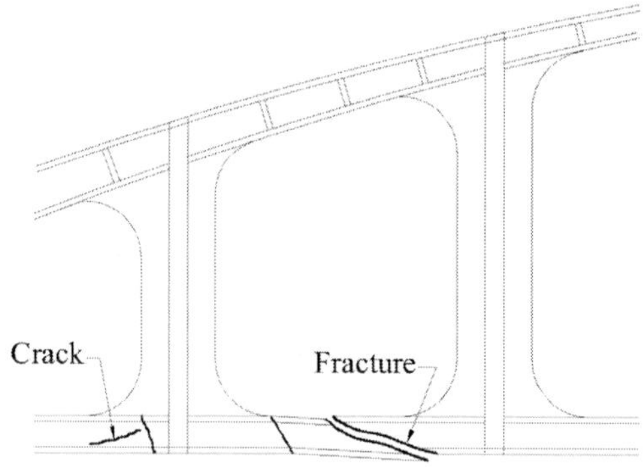

Detail 4

Kaulille Bridge Failure, Belgium, 1940

Figure 1.9f

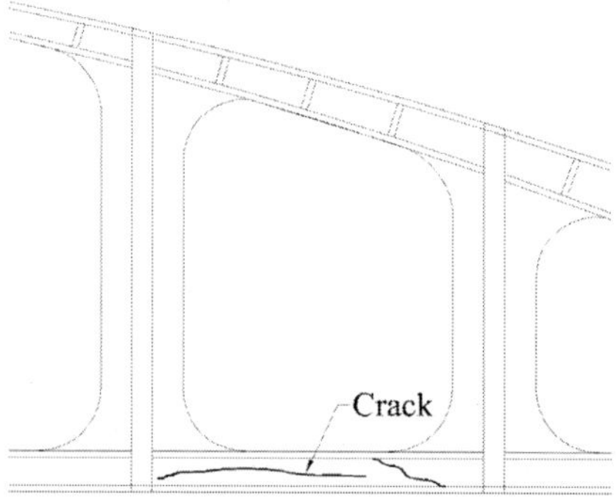

Detail 5

Kaulille Bridge Failure, Belgium, 1940

Figure 1.9g

1.2.2 Liberty Ships

The well-known failures of Liberty Ships started occurring in the winters of 1942-43 and 1943-44. There were two types of Liberty ships; vessels that carried general cargo and "T2" oil tankers. The all welded ships were built remarkably quickly (as little as 41 days) and the first ships were placed in service towards the end of 1941. Out of approximately 5000 merchant ships built during World War II, around 1000 ships experienced a total of approximately 1300 structural failures before April 1946. In 250 ships, serious failures, including complete fracture of deck and bottom hull plating occurred. Twenty ships either broke in two or were abandoned. One T2 tanker suddenly fractured in two on January 16, 1943 while lying in the outfitting dock. Stresses at the crown of the deck were calculated to be only 62 MPa (9,000 psi). Another T2 tanker broke in two in March 1943 at the entrance to New York Harbor where the sea was calm. The stresses at the crown were estimated to be about only 84 MPa (12,200 psi). Approximately 50% of the failures occurred by fracturing at square hatch corners, cut outs, and other surface discontinuities. About 40% of the failures started from weld defects including weld cracks, undercuts and lack of fusion. About 10% of the failures derived from metallurgical defects in the heat affected zones and notches. It was found that the steel plates had Charpy Vee Notch toughness of 20 Joules (15 ft.lb.) at 4° C (40°F) resulting in them being brittle in the North Atlantic during winter and spring. Figure 1.10, drawn from a photograph, shows a fractured Liberty Ship. [Lancaster (1992), Lancaster (1996), Gale Research Inc. (1994), Masubuchi (1980), Parker (1957)]. Improvements were made on ships

Fractured Liberty Ship, circa 1943

Figure 1.10

constructed during the latter part of World War II. The ships, so called "Victory" ships, were made with rounded hatch corners to reduce stress concentrations and performed very well [Grover (1954)].

1.2.3 Tanker Ponagansett, Boston Harbor

The tanker Ponagansett broke in two in Boston Harbor in December 1947. The fracture initiated from a tack weld between a small clip and the deck plate. The temperature at failure was 2° C (35°F) below the 10° C (50°F) Nil Ductility (see Chapter 2, 2.2.4). Subsequent improvements in design, materials and fabrication in ships did reduce the number of brittle failures. However, between 1951 and 1954, two new welded cargo ships and a welded tanker also broke in two [Masubuchi (1980)].

1.2.4 Kings Bridge, Melbourne, Australia

The 1962 collapse of the Kings Bridge over the Yarra River in Melbourne, Australia was caused by the cracking of transverse welds at the end of bottom flange cover plates when a single 47-ton trailer drove onto the bridge during a cold night. The failure can be seen in Figure 1.11, which is based upon a photograph that can be found in Boyd (1970). This led to the failure of four

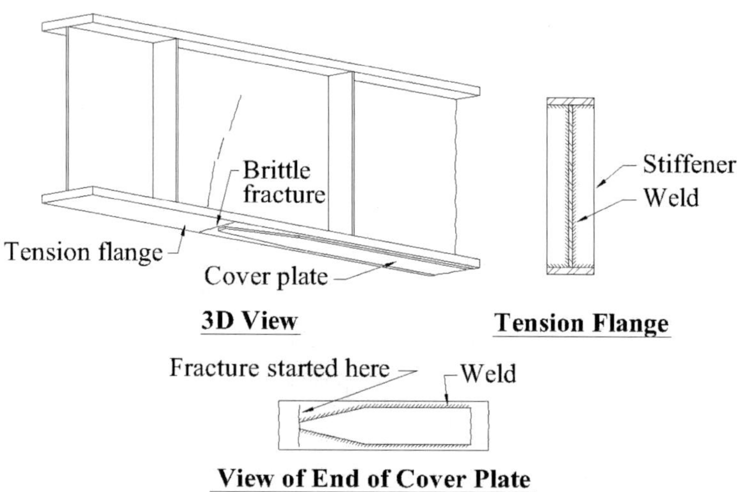

Details of Fractured Welded Plate Girder, King's Bridge Melbourne, Australia, 1962

Figure 1.11

girders with cracks extending to various depths in the webs of the girders and the collapse of one span. Stresses of approximately 138 MPa (20,000 p.s.i.) in the steel were estimated to have occurred when the trailer went onto the bridge. Welding of high strength steel, previously untried in Australia, was used and there was evidence that there was significant build up of residual stresses with some of the cracking occurring during fabrication [Gale Research Inc. (1994)].

1.2.5 Sea Gem, Offshore Platform, North Sea

The 1965 Sea Gem Offshore Platform disaster in the North Sea, during jacking operations, was attributed to the brittle failure of tie bars at the sharp radius corners (see Figures 1.12a and 1.12b). The vessel was originally built in 1953 for an aerial tramway for the U.S. Transportation Corps. Having been used during 1955-56, it was mothballed until it was reconverted as a jack up platform. The possibility of brittle fracture was considered and a Charpy Vee Notch toughness of 34 Joules (25 ft.lb.) at 0°C (32°F) was established as the requirement. Steel not meeting this requirement was normalized. However, this requirement was not considered necessary for the tie bars. The accident occurred in December 1965 during jacking operations when the air temperature was 3°C (37°F) and the water temperature about 6°C (43°F). Jacks, adopting grippers, were used to raise the platform and the platform was suspended from the jacks by tiebars (four per jack). Several tie bars fractured

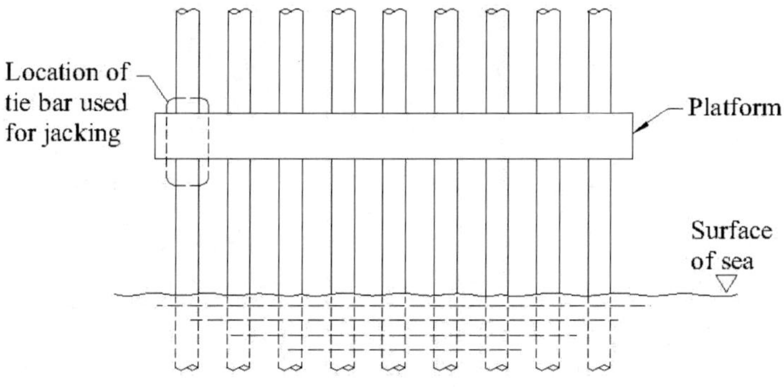

Sea Gem Disaster, North Sea, 1965
[Reproduced from Lancaster (1996) with permission from Woodhead Publishing Limited]

Figure 1.12a

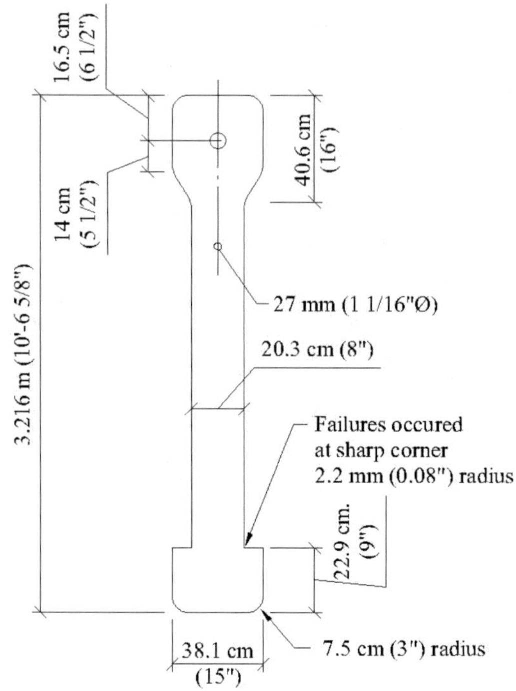

Tie Bar Detail

Sea Gem Disaster, North Sea, 1965
[Reproduced from Lancaster (1996) with permission from Woodhead Publishing Limited]

Figure 1.12b

and the platform fell horizontally. A crack in the bottom of the platform allowed water to pour in and the platform eventually capsized with many lives lost. The tie bar steel was made to ASTM A36 with no special requirement for notch ductility. The air temperature of 3°C (37°F) was low enough for the steel to be brittle (below the nil ductility temperature). The tie bars were flame cut and gouges were repaired by welding. These welds, which caused brittleness in the heat affected zones, also contributed to the failure [Lancaster (1992), Lancaster (1996)].

1.2.6 College of Science Building, Brooklyn, New York

Several steel girders failed at the College of Science Building, Brooklyn, New York in December 1971 while supporting five floors during construction of the roof. The welded steel-plate girders, which occurred above the first story, each supported a column at the end of its cantilever and another column along the backspan. The girders were unusual in that a portion of the length of the girders, adjacent to a support column, was open web to allow for ducts etc. Fracture occurred at the welded connection of a 38 mm (1-1/2 inch) thick diagonal plate and the 51 mm (2 inch) thick top flange. Figure 1.13, drawn from photographs that can be found in Kaminetzky (1991), illustrates the

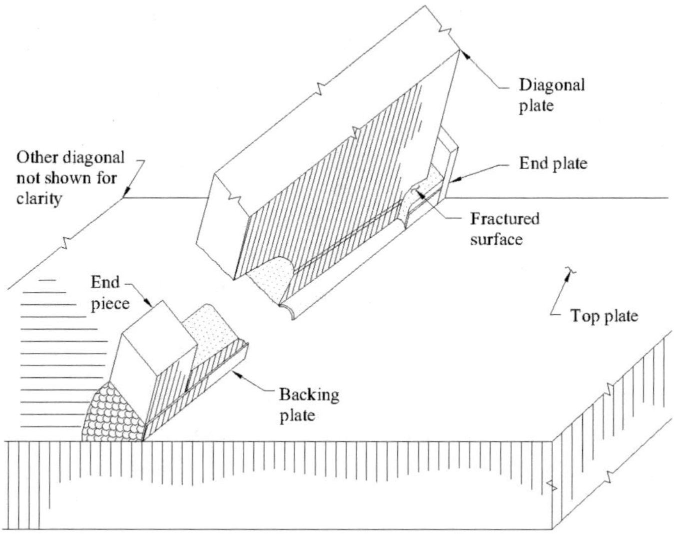

3D View of Fracture

College of Life Science Building, Brooklyn, New York, 1971

Figure 1.13

failure. Failure was attributed to lamellar tearing, undersized and defective welds, toe or underbead cracking and low fracture toughness of the high strength ASTM A441 steel. The building was not closed in and the temperature was cold which may have assisted in causing the fracture to be brittle. [Kaminetzky (1991)].

1.2.7 Atlantic Richfield Plaza Building, Los Angeles

Lamellar tearing was encountered in beam column connections at the twin towers of the Atlantic Richfield Plaza Building in Los Angeles in 1973. The cracks appeared in the base metal and within the welds. Approximately 10% of beam flange to column connections, which had flange thickness exceeding 38 mm (1-1/2 inches), were affected [Kaminetzky (1991)].

1.2.8 Alexander Keilland, Offshore Platform, North Sea

The Alexander Keilland offshore platform disaster in the North Sea in 1980 was a major example of potential problems with offshore structures when 123 people died [Lancaster (1996), McEvily (2001), Norwegian Institute of Technology]. The event, involving failure of a brace, occurred during a storm with wave heights of 6 m (20 feet) to 8 m (26 feet). Failure of the cylindrical brace initiated at a fillet weld for a hydrophone penetration through a brace. The brace failed as a result of a crack propagating circumferentially around the brace adjacent to the hydrophone (an electronic device that receives signals from the well cap located on the sea floor). Figures 1.14a and 1.14b illustrate

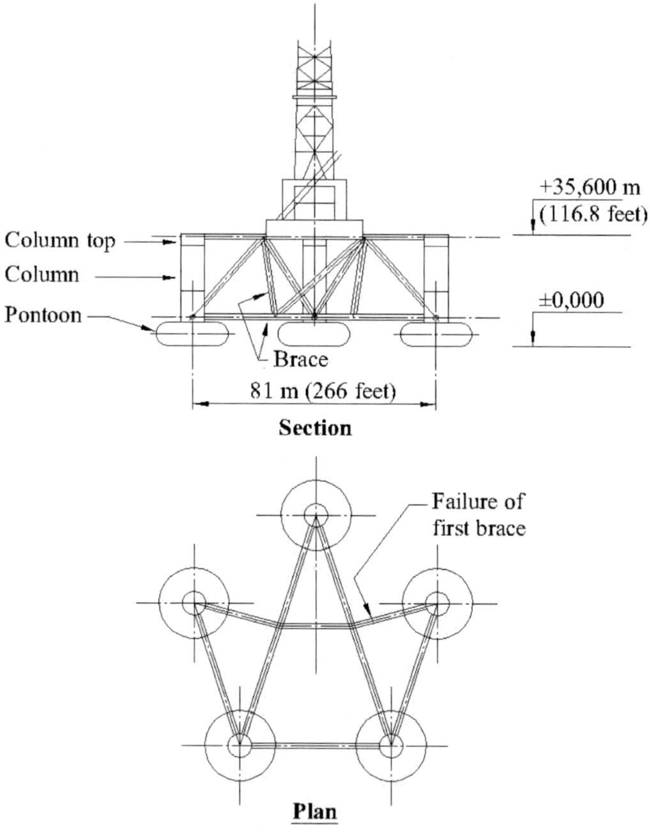

Alexander Keilland Accident, North Sea, 1980
[From McEvily (2001), reproduced with permission of John Wiley and Sons, Inc.]

Figure 1.14a

the structure and Figure 1.14c shows the fracture propagating from the hydrophone. Subsequently, other braces failed, due to significant increase in stresses, and column D broke away (see Figure 1.14b). The platform heeled over about 35° then slowly capsized. There was evidence that there were cracks, as long as 75mm (3 inches), in the weld prior to the disaster. The initial cracks may have occurred due to thermal strains resulting from welding. Also, the hydrophone steel material had low strength and ductility and was susceptible to lamellar tearing.

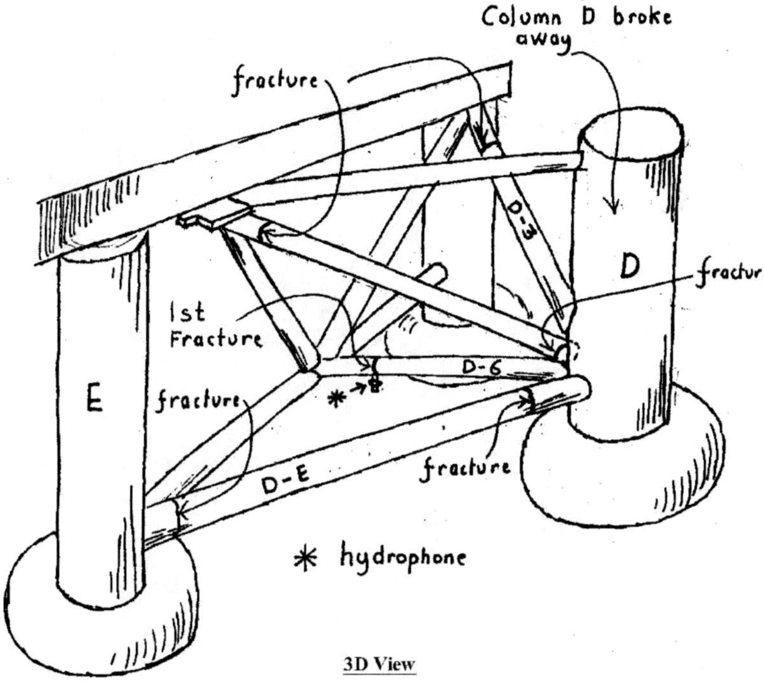

Alexander Keilland Accident, North Sea, 1980
[From McEvily (2001), reproduced with permission of John Wiley and Sons, Inc.]

Figure 1.14b

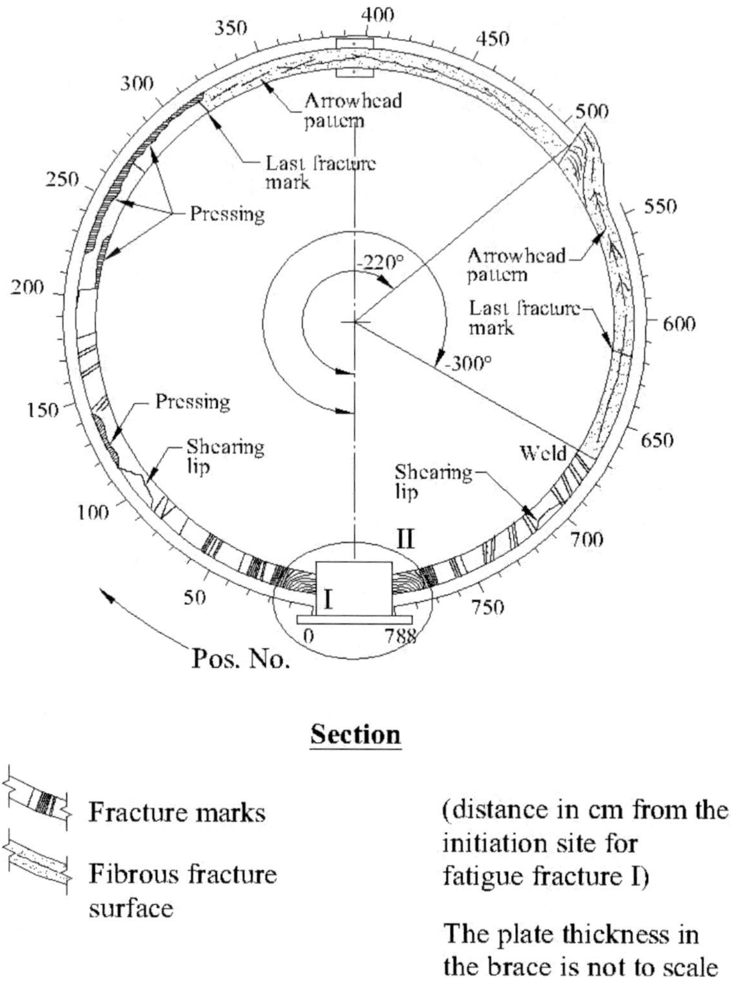

Fracture Surface Features of the Failed Brace D-6.
Note Two Fatigue Origins at Locations I
and Ii at the Welded in Hydrophone.

Alexander Keilland Accident, North Sea, 1980
[From McEvily (2001), reproduced with permission of John Wiley and Sons, Inc.]

Figure 1.14c

1.2.9 Wolftrap Center, Fairfax, Virginia

A large crack occurred in one of the main roof girders at the Wolftrap Center, Fairfax, Virginia, which fortunately did not collapse due to sufficient redundancy permitting alternate paths of load transfer. Figure 1.15, which was drawn from a photograph that can be found in Kaminetzky (1991), illustrates the failure. The crack was as wide as 38 mm (1½ inches). The girder was made from welded steel plates and stress levels were below the allowable design stresses for dead and live loads. Close examination revealed the failure initiated from a large flaw at a discontinuous back up plate of a full penetration weld [Kaminetzky (1991)]. Also, there were no minimum toughness requirements for the girders.

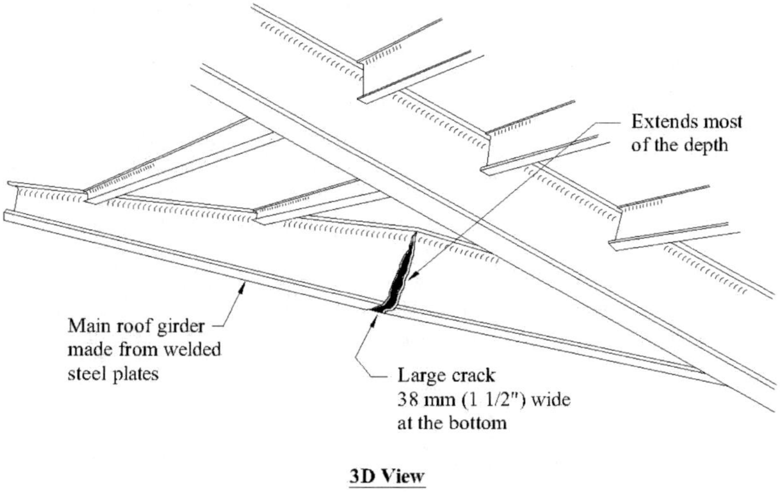

Wolftrap Center Failure, Fairfax, Virginia

Figure 1.15

1.2.10 Offshore Structures

According to Barsom and Rolphe (1999), cracking occurred in 69 vessels over 10,000 gross tons between 1984 and 1988. It was observed that a disproportionate amount of cracking occurred in the vessels made with high tensile steel compared with the vessels made with mild steel. The cracks, at fatigue sensitive details, were due to fairly severe wave forces leading to fatigue crack initiation followed by propagation.

Although the cracks were serious, they did not lead to complete failures. A thorough investigation using a fracture mechanics methodology led to improvements in inspection, details, welding and the use of holes for crack arresting.

According to Lancaster (1996), there have been several cases of fatigue failures on offshore structures. Some of these were due to weld root defects, stress discontinuities, and discontinuity of back-up strips.

1.2.11 1994 Northridge Earthquake, California

Damage to a significant number (over 100) of welded steel moment frame buildings, including cracks in welds some of which propagated into the columns, were found in Los Angeles, California, following the January 17, 1994 Northridge Earthquake [Bertero et al (1994), Bonowitz and Youssef (1995), FEMA 267 (1995), FEMA 350 (2000), 351 (2000), 352 (2000) and Maranian (1997)].

A special research and development project, funded by FEMA, called the SAC Project, was carried out to investigate the failures and develop improvements in steel moment frame connections.

Figure 1.16a indicates the typical pre-Northridge moment frame connection with the top and bottom flanges of the beam connected to the column flange with complete penetration welds using back up bars. The webs of the beams were connected with bolts to shear tabs, which were welded both sides with fillet welds to the columns.

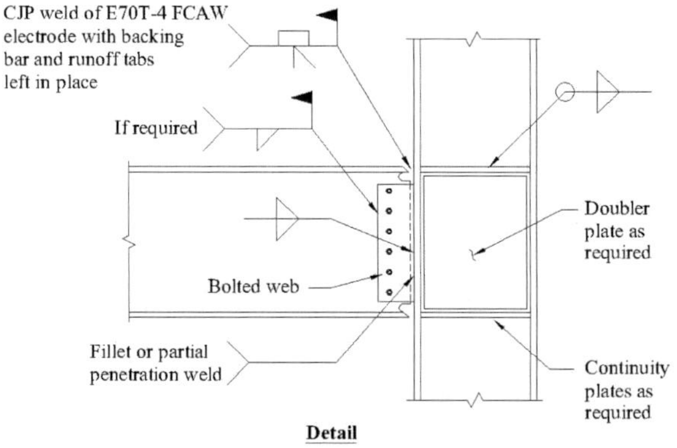

Typical Pre-Northridge Connection

Figure 1.16a

Figure 1.16b shows a column, which cracked, all the way through the thick flange section and into the web. The column was the end column of a moment frame in an eleven-story building. The fracture occurred at the second floor. The column, due to the cyclic loading, jumped, as there was a distinct step in the column at the failure line. Also, note the spall at the inside face of the flange adjacent to the web. It is also interesting to note that a second building of the same design, located in the same vicinity, had a similar fracture at the same column at the same floor.

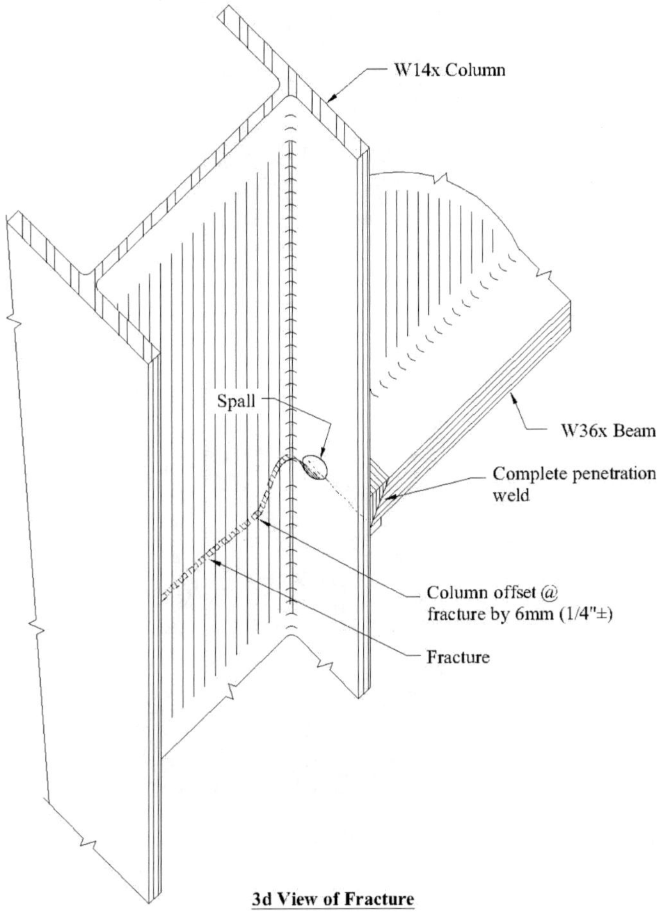

Fractured End Column at Second Level of a Moment Frame from an 11-Story Building, Northridge Earthquake, Los Angeles, 1994

Figure 1.16b

Figure 1.16c shows another column that cracked all the way through. This specimen displays a typical brittle fracture with little or no evidence of ductility in the fracture zone.

Pieces together

Pieces separated

Column Flange Specimen Extracted from Two-story Building, Northridge Earthquake, Los Angeles, 1994
[*Photo* courtesy of Rafael Franco]

Figure 1.16c

Figure 1.16d shows a column cracked in four places. The fractured column occurred at the second level of the same eleven-story building where the fractured end column, described above and shown in Figure 1.16b, had occurred. Two of the cracks occurred away from welds. The specimen was given to the SAC project. The reasons for the fractures, according to the writer's knowledge, were not reported.

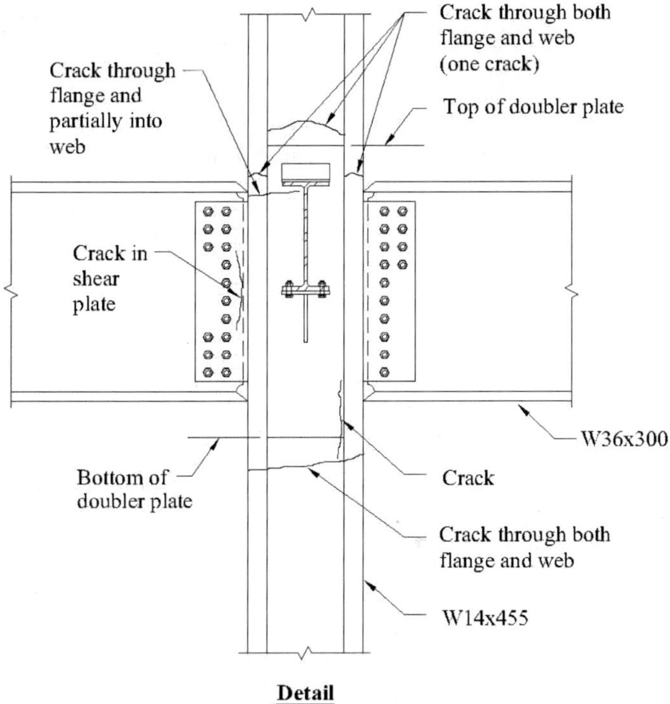

Damage at Third Floor Joint of a Three-bay Frame from an 11-story Building, Northridge Earthquake, Los Angeles, 1994

Figure 1.16d

Various causes for the cracking have been identified by many researchers. These reasons include low cycle fatigue, low notch toughness in the weld metal, poor welding procedures adopting high heat input, stress risers at the back-up bar, and significant stress concentrations particularly at the weld access hole in the web. Richard et al (1995) found significant stress and strain concentrations in the Pre-Northridge connection associated with the tendency of the flanges to also accommodate significant shear.

There were many tests carried out on moment frame connections for the SAC project and for prototypes used on projects following the Northridge Earthquake. FEMA 355D (2000) on Connection Performance summarizes tests carried out in the SAC Project on Pre-Northridge (earthquake) and Post-Northridge beam to column connections. There are also reports by the researchers who carried out research and testing for the SAC project. Although not mentioned in FEMA 350, some researchers discussed low cycle fatigue associated with connection performance [Ricles et al (2000)]. Partridge et al (2000), independent from the SAC project, cited low cycle fatigue as a major issue with beam to column seismic moment connections. Some of the Post-Northridge beam to column connections prematurely failed in a brittle manner. One test (not for the SAC project), involving a cover plate applied at each flange of a W36x359 moment frame girder welded to a W36x670 column, prematurely failed when yielding of the girder was at approximately 8 to 10 times the yield strain and total rotation was approximately 1%. (see Figures 1.16e, 1.16f and 1.16g). The cover plate detail and the failure appeared similar

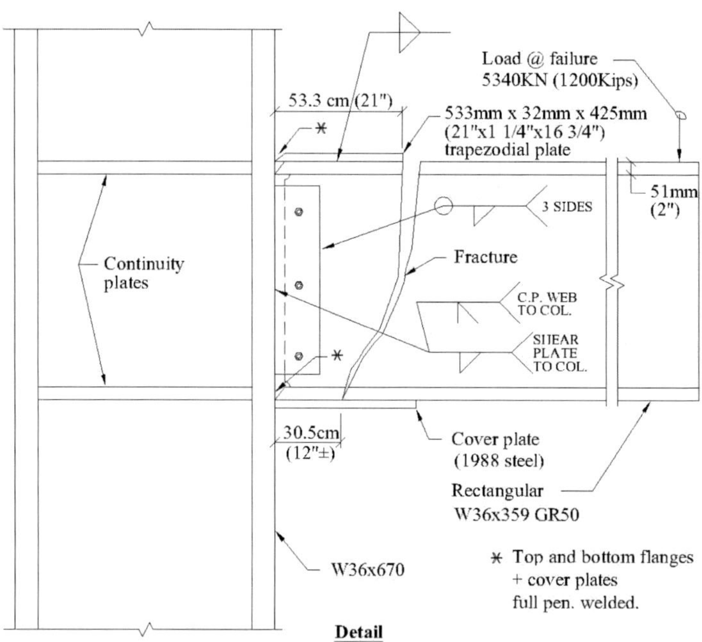

Post-Northridge Beam-to-Column Test

Figure 1.16e

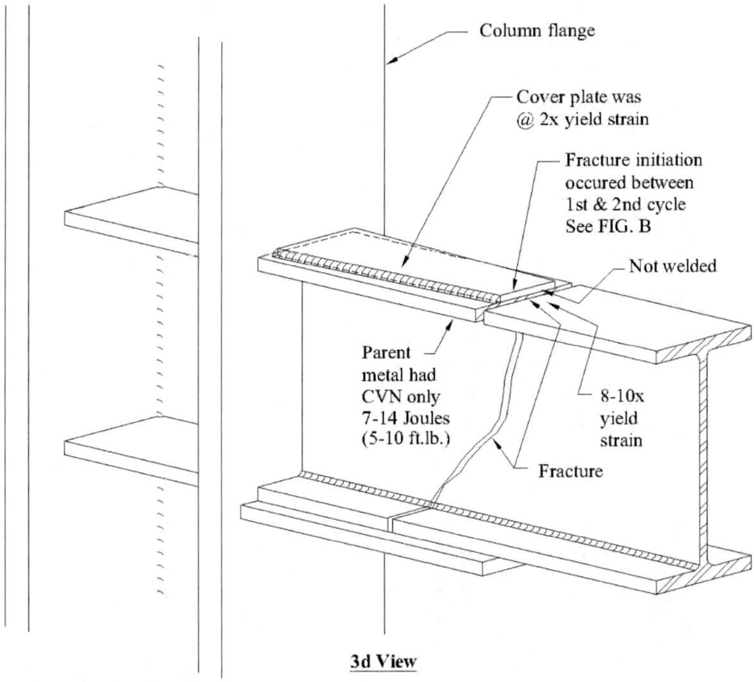

Post-Northridge Beam-to-Column Test

Figure 1.16f

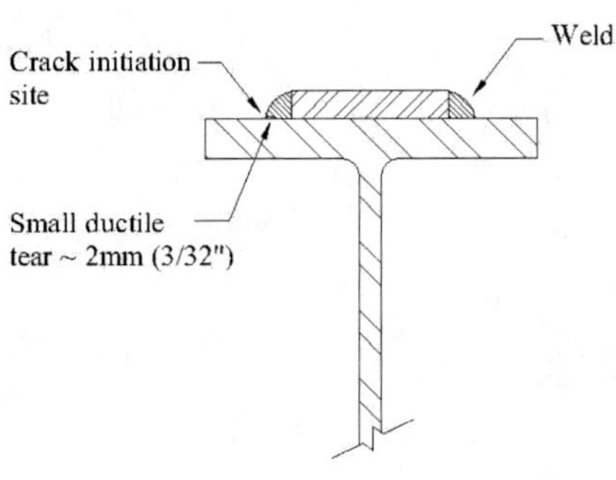

Post-Northridge Beam-to-Column Test

Figure 1.16g

to the Kings Bridge failure (see Section 1.2.4). In another independent testing program, one of six beam-to-column moment connections using cover plates, carried out circa 1995, failed in a brittle manner. The test specimen comprised a W36 x 150 beam connected to a W14 x 426 column and both the cover plates and beam flanges were welded with complete penetration welds. Fractrographic examinations were carried out by Harrison and Webster [Harrison and Webster 1995]. According to Harrison and Webster, the crack propagated in a brittle manner from the central region of the welded connection in the absence of a pre-existing defect on the fracture face, was then arrested in the heat affected zone [see Chapter 4, Section 4.1.4 (iii)] and then re-extended in a ductile manner. Harrison and Webster noted the need to have sufficient toughness in the heat-affected zone to address defects in this area.

Although less reported, there were incidences involving significant damage to braced frames. Damage included weld failures at gusset plates (See Figure 1.17). Damage to braces other than welds was discussed in Section 1.1.9.

Fracture of Steel Tube Brace Member at Gusset Plate
Northridge Earthquake, Los Angeles, 1994

Figure 1.17

1.2.12 1995 Hyogoken-Nanbu (Kobe) Earthquake, Japan

Significant damage occurred in steel buildings in the January 17, 1995 Hyogoken-Nanbu (Kobe) earthquake (AIJ (1995)). Most of the damage occurred in low to medium rise buildings. Damage occurred in steel moment frames, of a similar nature to that observed after the 1994 Northridge earthquake. Figure 1.18a indicates the typical moment frame connection used.

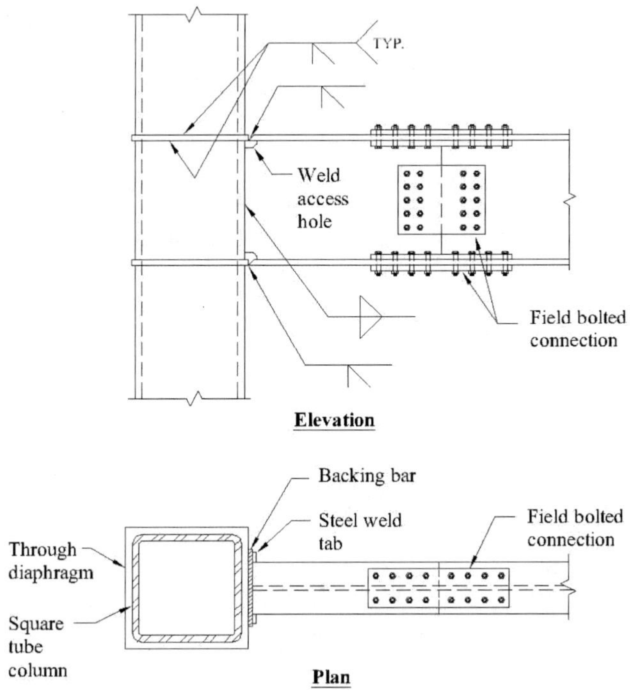

Typical Moment Connection Used in Japan

Figure 1.18a

A portion of beam was shop welded to continuity plates through the square tube column. The box columns were typically made from two H sections welded together with the outstanding flanges removed. The "tree" was then bolted in the field using flange and web plates. Damage included complete fracture of beam flange and web welds at columns, complete brittle fracture of large welded steel tube columns typically below the beam flange and adjacent to weld splices (see Figures 1.18b and 1.18c). The figures were drawn from photographs that can be found in AIJ (1995).

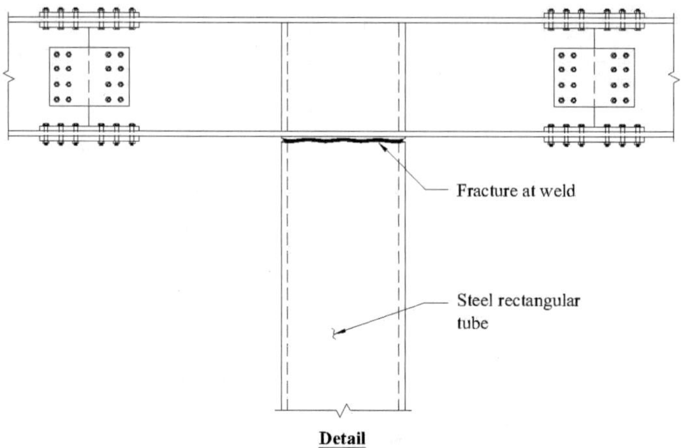

Failure of Weld at Top of Column
Hyogoken-Nanbu (Kobe) Earthquake, Japan, 1995

Figure 1.18b

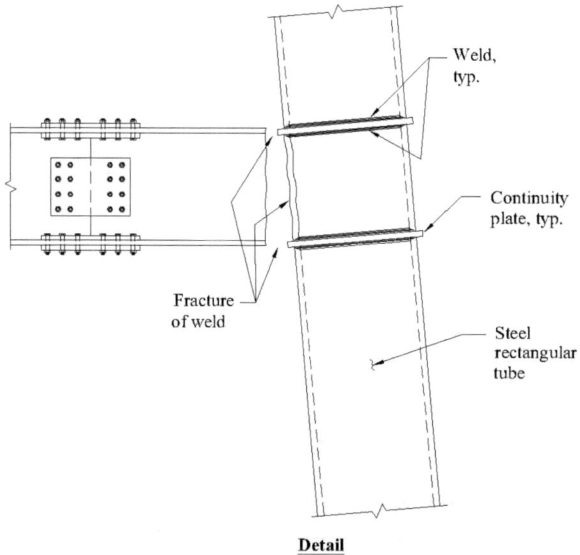

Failure at Beam-to-Column Connection
Hyogoken-Nanbu (Kobe) Earthquake, Japan, 1995

Figure 1.18c

1.2.13 Fatigue Cracks in Welded Steel Bridges

Fatigue cracks in several steel bridges have been reported by Fisher et al (1998). Cracking in the welds at the ends of cover plates on steel beams (similar to the Kings Bridge, Section 1.2.4), with cover plate thicknesses greater than 19 mm (3/4 inch) have occurred and were found to have a low fatigue rating. Significant stress concentrations at the abrupt change in section caused small cracks to occur at the toe of fillet welds. These small cracks combined to form one large crack.

Fatigue cracks have also been found emanating from discontinuities in horizontal welded gusset plates where rectangular holes have been made for vertical stiffeners. A vertical crack was discovered in a fascia girder of a nine girder four span bridge on a U.S.A. interstate highway approximately nine years after the bridge was built [Fisher et al (1998), Fisher et al (1997)]. The crack initiated at an unfused groove weld required to connect horizontal stiffener plates. The crack extended upwards to about mid-depth of the girder and downwards penetrating the bottom flange (see Figure 1.19). Later on, a second crack was found in the same girder also commencing at the unfused groove weld in the longitudinal stiffener. Figure 1.19 was drawn from a photograph that can be found in Fisher et al (1998).

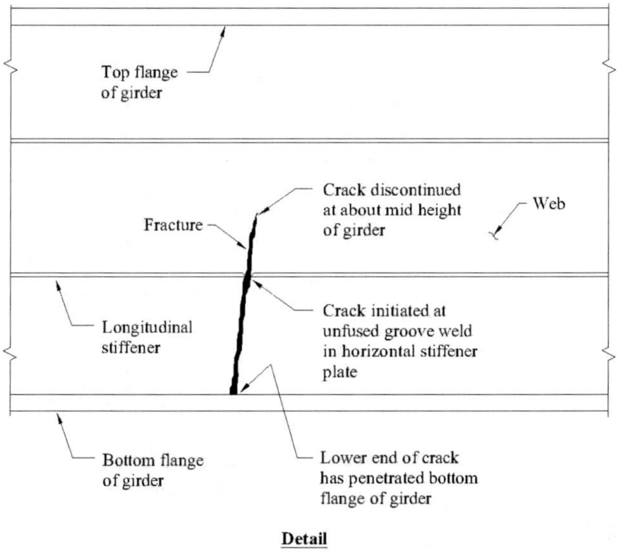

Fracture of Interstate Highway Bridge Girder

Figure 1.19

RECOMMENDED READING - CHAPTER 1

Architectural Institute of Japan (AIJ) (1995); Preliminary Reconnaissance Report of the 1995 Hyogoken-Nanbu Earthquake, English Edition, April.

Barsom, J.M. and Rolphe, S.T., (1999); Fracture and Fatigue Control in Structures, 3rd Edition ASTM.

Bertero, V.V., Anderson, J.C. and Krawinkler, H., (1994); 'Performance of Steel Building Structures During the Northridge Earthquake' College of Engineering, U. C. Berkeley.

Boyd, G.M., (1970); Brittle Fracture in Steel Structures. Butterworth, London.

Campus F., (1952); Effects of Residual Stresses on the Behavior of Structures; F. Campus.

Fisher, J.W., Kulak, G.L., Smith, I.F.C., (1998); A Fatigue Primer for Structural Engineers; National Steel Bridge Alliance.

Fisher, J.W., Kulak, G.L., Smith, I.F.C., (1997); A Fatigue Primer for Strutural Engineers; Advanced Technology for Large Structural Systems (ATSS) Report No. 97-11, October.

Gale Research Inc. (1994); When Technology Fails, Significant Technological Disasters Accidents and Failures of the Twentieth Century; Edited by Neil Schlager.

Kaminetzky, D., (1991); Design and Construction Failures, Lessons from Forensic Investigations; McGraw-Hill, Inc.

Lancaster, J., (1996); Engineering Catastrophes, Causes and Effects of Major Accidents; John Lancaster; Abington Publishing.

Levy, M. and Salvadori, M., (1992); Why Buildings Fall Down; ISBN 0-393-03356-2, W.W. Norton & Company.

Masubuchi, K., (1980); Analysis of Welded Structures, MIT, Pergamon Press.

McEvily, A.J., (2001); Metal Failures; Mechanisms Analysis, Prevention, Wiley-Interscience.

Parker, E.R. (1957); Brittle Behavior of Engineering Structures; Published by John Wiley & Sons Inc.

CHAPTER 2

FRACTURE AND FATIGUE

2.1 INTRODUCTION

Fracture mechanics was first developed by A.A. Griffith in the 1920s from his studies on the fracture behavior of silica glass. His theories were based upon the first law of thermodynamics, which states that, in a closed system, energy is conserved. Griffith demonstrated why the fracture strength of silica glass was substantially lower than the theoretical strength due to the existence of small defects. Griffith considered that crack propagation would occur when the rate of strain energy release equates to the rate at which energy is being taken up from the occurrence of an increase in the crack surface.

Further development was carried out by G.R. Irwin in 1948 and E. Orowan in 1955 on the fracture of steel. In their dissertations, they also considered the plastic work done during fracture. George Irwin was recognized as the father of Fracture Mechanics.

High cycle fatigue has been quite well understood and researched over many decades particularly with regard to bridges. However, the understanding of the effects of defects, low temperature, high strain rates, constraint conditions and other phenomena is, in general, not well understood by practicing engineers.

The consideration of non-ductile modes of failure in steel framed buildings is not thoroughly addressed in the present standards of practice regarding building design. Of particular concern is the occurrence of short term cyclic loading that may induce large (plastic) deformation in a building frame such as earthquakes (see Chapter 1 for examples of failures). In an effort to confront these issues, the following discussion has been developed. The discussion is intended to be a general overview of elementary principles related to fracture and fatigue. The overview is not rigorous and does not completely explain all of the underlying theoretical concerns in an effort to provide a concise but useful presentation of the relevant topics. Several of the figures shown are similar to figures typically found in books on fracture mechanics. Further reading of Fracture Mechanics books is recommended and a bibliography is provided at the end of this section. Of particular note, regarding Fracture and Fatigue, is the reference by Dowling (1999 and 2007). This text provides a thorough introduction to the concepts discussed below. Other excellent references are Barsom and Rolphe (1999) and McEvily (2001).

2.2 FRACTURE

2.2.1 Types of Fractures

Essentially two types of fracture can occur, **Transgranular** (see Figure 2.1a) and **Intergranular** (see Figure 2.1b). Transgranular type failure is more common under cyclic loading.

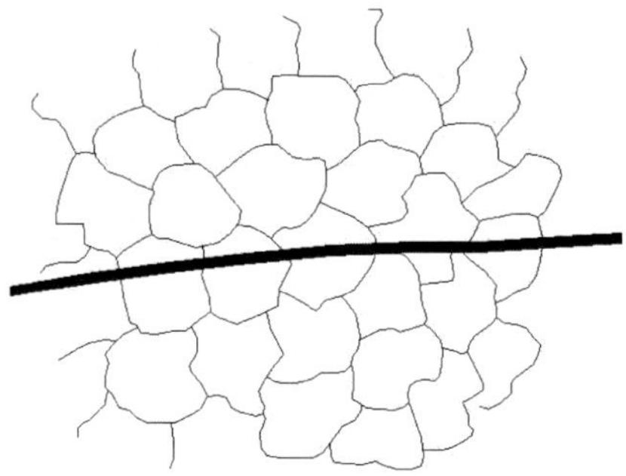

Transgranular Facture
[Reproduced from "Analysis of Welded Structures" by Koichi Masabuchi, Pergamon Press, 1980 with permission of Dr. K. Masabuchi]

Figure 2.1a

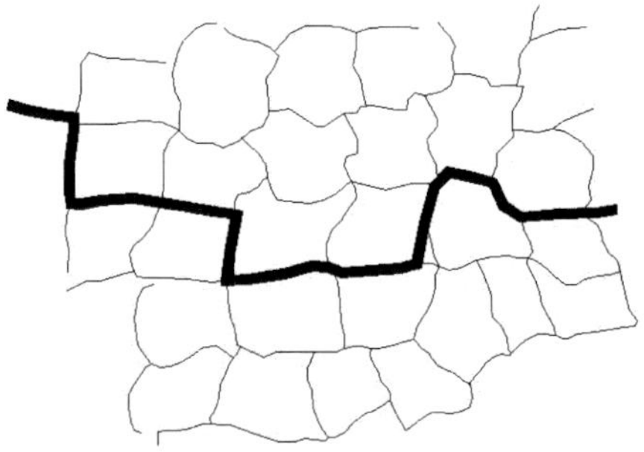

Intergranular Facture
[Reproduced from "Analysis of Welded Structures" by Koichi Masabuchi, Pergamon Press, 1980 with permission of Dr. K. Masabuchi]

Figure 2.1b

A separation process known as cleavage occurs in Transgranular brittle fractures in steel. Chevron markings, which are macroscopic tear lines, can be seen on the fracture surface of Transgranular brittle failures. Bright reflections can be observed by eye and on a microscopic level. They show up as tear lines. The brightness is an indication of lack of ductility. Figure 1.16c, showing column fracture found in a building following the 1994 Northridge Earthquake, is an example.

The tear lines, visible at a microscopic level, create a pattern known as river pattern like tributaries merging into a mainstream. The tear lines, which tend to be perpendicular to the crack front, dissipate energy of the material as they exert a drag on the crack propagation. The river pattern helps locate the origin of the brittle fracture. Cracks in steel, which are brittle, occur at high velocity and strain rate.

Integranular failures occur due to weakness of the grain boundaries usually due to the presence of impurity elements at the grain boundaries. This failure surface has a bright "rock candy" appearance at low magnification, with the absence of cleavage type fracture patterns.

Ductile failure of steel occurs due to plastic shear deformation and is absent of a crystalline, bright, appearance. Thus it is much duller and smoother in appearance when compared to a brittle fracture.

2.2.2 Fractures of Flawed Members

If a sharp flaw is present in a structural component, non-ductile failure may occur even if the material is capable of large plastic deformation. The application of load can cause high stress and strain concentrations adjacent to the flaw which may be sufficient to overcome the internal strain energy capacity, in a similar manner to splitting a piece of wood with an axe. The principles of *fracture mechanics*, the study of cracks in solids, must be invoked to analyze these situations. The resistance of a material to fracture is generally quantified with the concept of *fracture toughness*. Fracture toughness is a material property derived from testing. In general, higher strength steels have lower fracture toughness and are more susceptible to fracture failure. The converse is true for lower strength steels. The concept of a *stress intensity factor* is used to assess how the presence of a crack affects the stress within a component. The stress intensity factor is quite similar in concept to the stress concentration factor associated with a "stress raiser" in a component as is often studied in undergraduate Strength-of-Materials courses. These concepts will be discussed in greater detail below.

The presence of a crack or crack-like flaw in a structural component may weaken it so that it fails in a non-ductile fashion. Examples of crack like flaws include cracks themselves, surface scratches, voids in welds and inclusions in the material matrix. As was mentioned earlier, these flaws act as stress raisers by distorting the uniform stress field in the vicinity of the flaw. Consider a linear elastic analysis, similar to an elliptical hole (see Figures 2.2a & 2.2b). As the radius of the "defect/crack tip" (flaw tip) becomes small, the theoretical stress at the crack tip (similar to an elliptical hole) becomes infinite (see Figure 2.2c). However an infinite stress cannot exist, so linear elastic analysis of the defect/crack tip stress field is not valid. For metals in general,

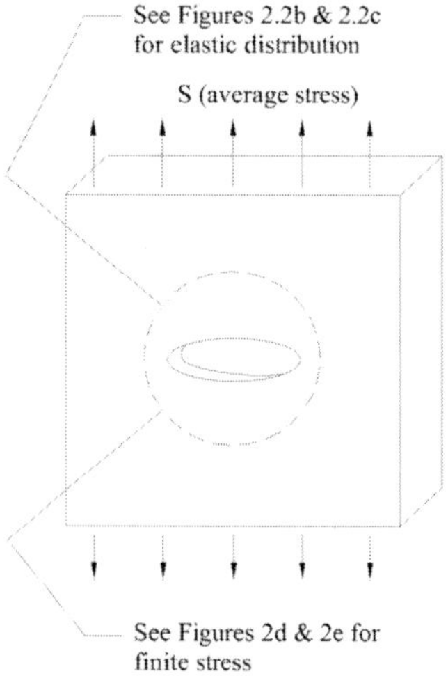

Elliptical Hole

Figure 2.2a

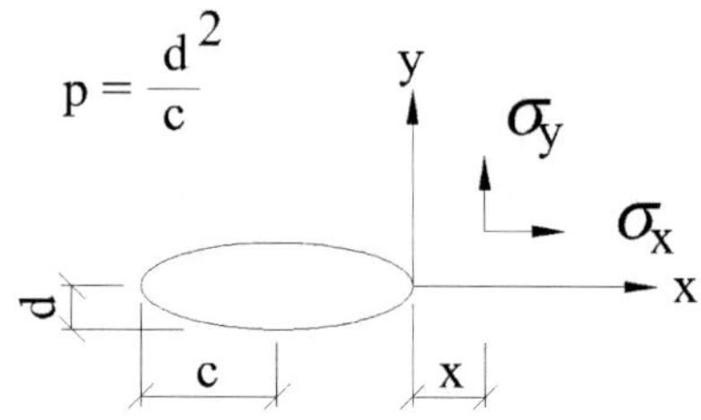

Elliptical Properties

Figure 2.2b

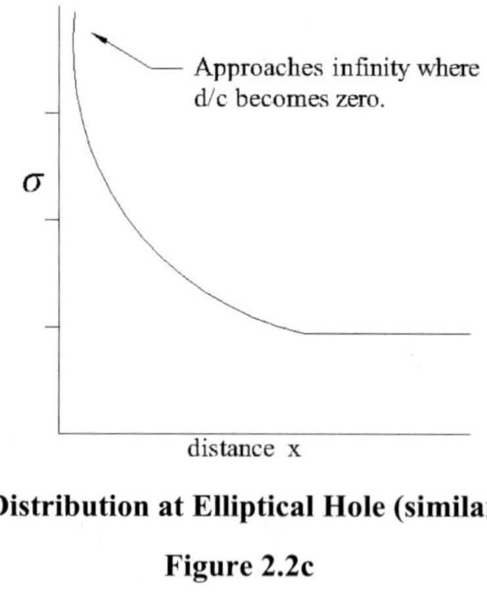

Elastic Stress Distribution at Elliptical Hole (similar defect/crack)

Figure 2.2c

an intense localized deformation occurs at the crack tip resulting in a localized plastic zone (see Figures 2.2d and 2.2e). The large stresses that would theoretically exist at the crack tip are redistributed over a larger region by plastic deformation in the vicinity of the crack. If an applied stress is large enough, the crack may suddenly grow causing a brittle failure. To characterize the severity of a crack, the concept of a stress intensity factor is used. The use of a stress intensity factor, commonly denoted as 'K', implies that the base material exhibits linearly elastic behavior. This approach is referred to as *linear elastic fracture mechanics* (LEFM).

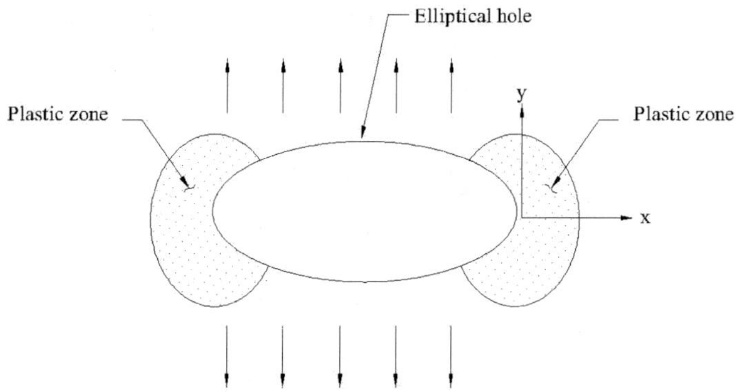

Finite Stress at Elliptical Hole (similar defect/crack)

Figure 2.2d

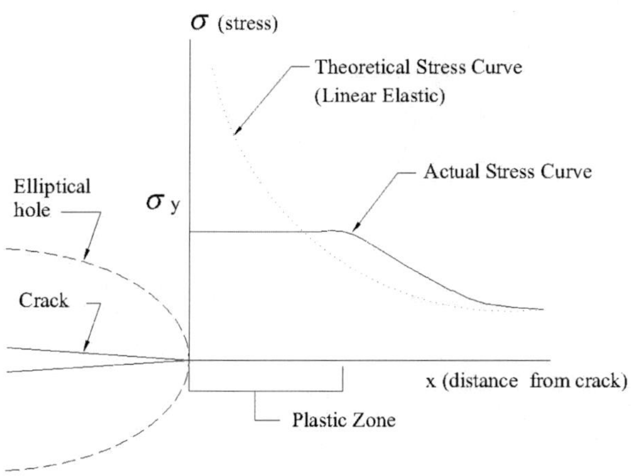

Finite Stress at Elliptical Hole (similar defect/crack)

Figure 2.2e

A component can resist a crack (or defect) without brittle failure as long as K is less than the fracture toughness of the material, K_c. Consider the center-cracked plate in Figure 2.3 as an example. The stress intensity factor K may be calculated as follows:

$$K = S\sqrt{\pi a}$$

Where:

S = uniform stress on the gross cross section of the plate
a = crack length

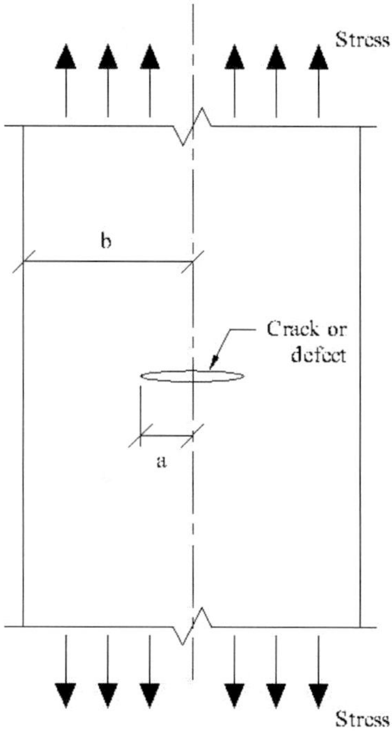

Geometry of Internal Crack (or defect)

Figure 2.3

This equation is similar to Euler Buckling column instability. Comparisons between Column Instability and Crack Instability can be seen in Figures 2.4a and 2.4b.

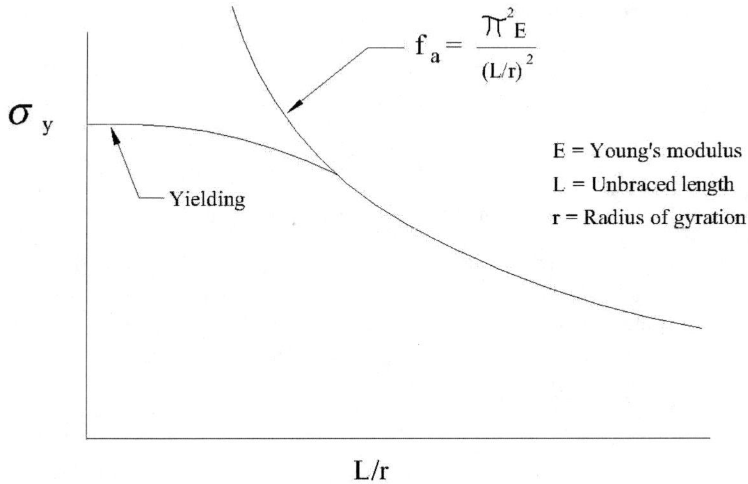

Column Instability

Figure 2.4a

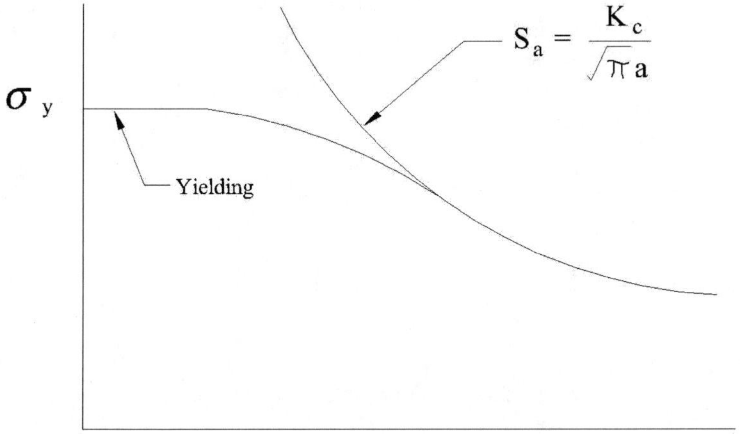

(See Section 2.2.2 for definitions)

Crack Instability

Figure 2.4b

Consequently, the allowable stress, S_a, on the gross cross section of the plate may be represented by:

$$S_a = \frac{K_c}{\sqrt{\pi a}}$$

Where:

K_c = fracture toughness of the material

Consider a situation where the allowable stress, S_a, calculated from the equation above equals the yield stress of the material, σ_y. The equation may be rearranged to describe the crack length at which this situation will occur.

$$a_t = \frac{1}{\pi}\left(\frac{K_c}{\sigma_y}\right)^2$$

Where:

a_t = the "transition" crack length

The transition crack length, a_t, may be considered the approximate demarcation between ductile and brittle behavior of the plate. In general, the impact of crack lengths greater than or equal to a_t will have to be considered using LEFM (with the exception of cracks that are large relative to the geometry of the cracked members, see Section 2.2.8). Crack lengths less than a_t will result in little or no strength reduction due to the crack. It should be noted that this portion of the discussion has been limited to a uniaxial loaded center cracked plate condition.

Member geometry and crack configuration are critical considerations in practical application of LEFM. A more practical form of the stress intensity factor equation is:

$$K = YS\sqrt{\pi a}$$

Where:

Y = function of member geometry and crack configuration

Provisions for determination of Y for various types of crack configurations and loading conditions may be found in the references for this section. Typical calibration curves for cracked plate geometries are shown in Figure 2.5.

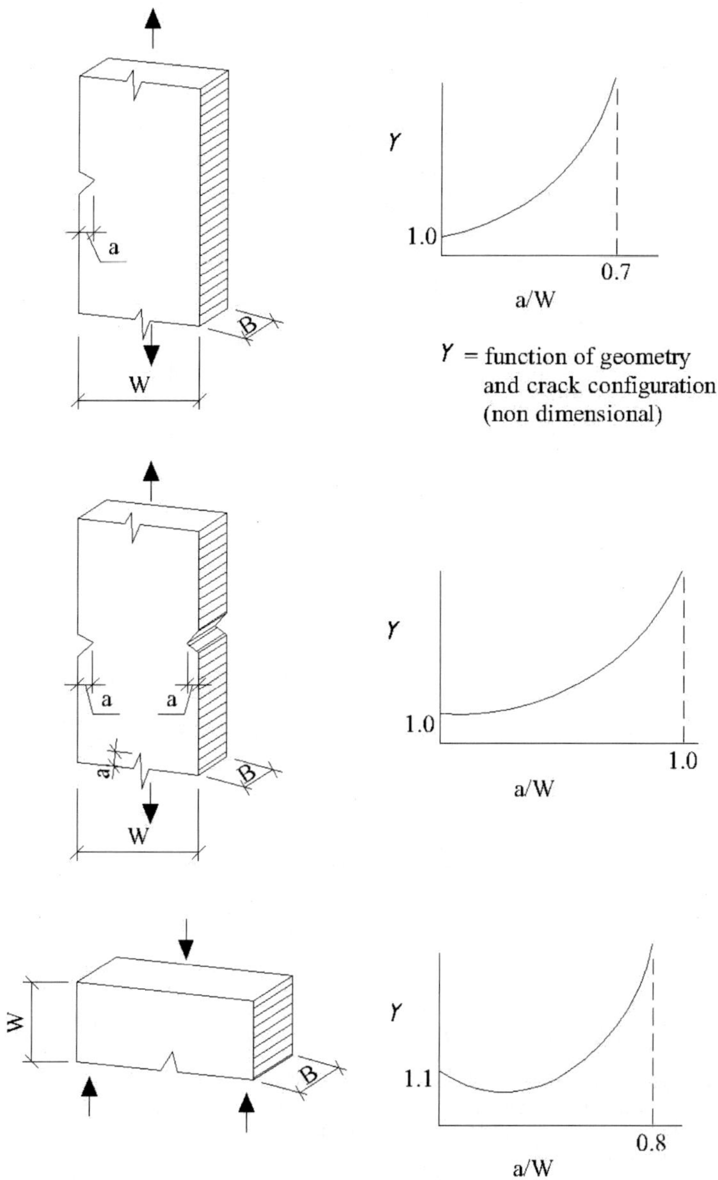

Calibration Curves for Cracked-Plate Geometries

Figure 2.5

2.2.3 Plane Stress and Plane Strain

Plane Stress

Essentially, for plane stress conditions, there is little or no restraint to Poisson's Ratio effect such that the transverse stress perpendicular to the principal stress is small. Upon increased application of force, necking occurs since transverse movement is mostly unrestrained. As a result, static tension failure is usually characterized by dislocation along the shear plane, at 45 degrees to the line of action of the force, which promotes ductile behavior. (See Figure 2.6) The actual process involves voids tending to nucleate around inclusions (typically carbides in mild steels) when subjected to tri-axial stress [Kanvinde and Deierlein (2005)]. Increase in the void size is highly dependent on the plastic strain. When the void growth demand exceeds the void growth capacity or critical void size, fracture occurs.

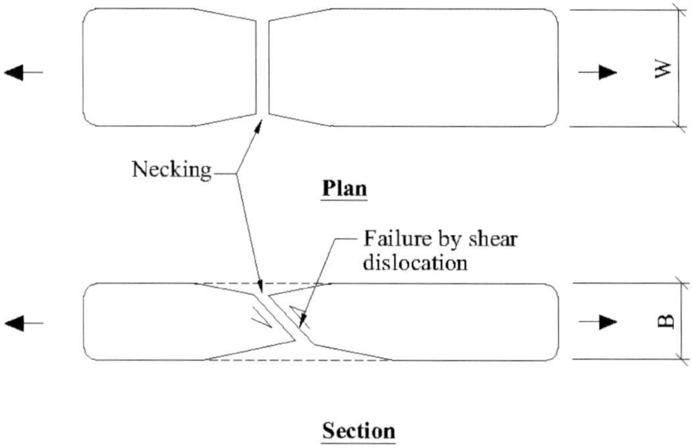

Ductile Failure in a Plane Stress Condition

Figure 2.6

Plane Strain

For plane strain conditions, Poissons Ratio effect is constrained. Therefore a significant tensile stress, transverse to the applied tensile stress, occurs and necking, which occurs in the plane stress condition, is prevented. Although the strength of the material is enhanced due to the constraining effects, yielding on through-thickness shear planes is not possible. This condition tends to promote a cleavage (brittle) failure in the center area with partial ductile failure at the edges as shown in Figure 2.7. Furthermore, the fracture

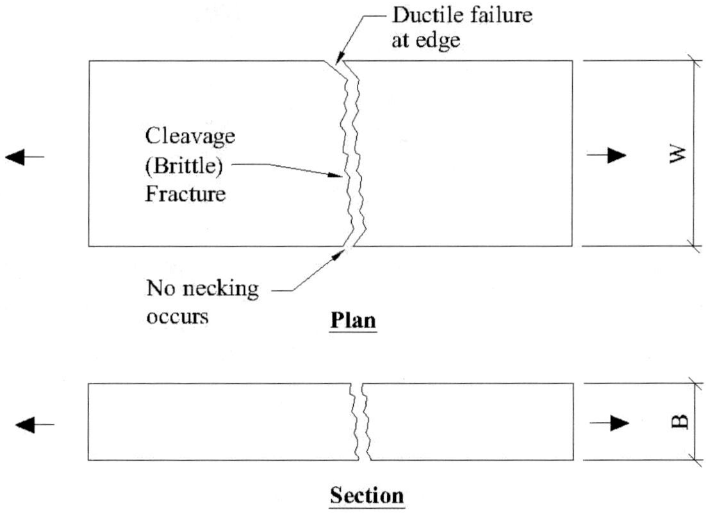

Brittle Failure in a Plane Strain Condition

Figure 2.7

toughness is significantly reduced. It has been shown that for a plate thickness greater than or equal to β, the fracture toughness of the specimen is not expected to decrease. The plane strain general relationship, considering fracture toughness, is as follows:

$$\beta \geq 2.5\left(\frac{K_{IC}}{\sigma_y}\right)^2$$

Where:

K_{IC} = critical stress intensity factor under conditions of maximum constraint and β is the thickness of the component

σ_y = static tensile yield strength obtained in slow tension test

The variation in fracture toughness with thickness is represented in Figure 2.8.

Fracture toughness of steel under constraint (plane strain condition) is significantly less than for an unconstrained condition (plane stress condition). Therefore, the value for fracture toughness, K_c, used in the above equations, is

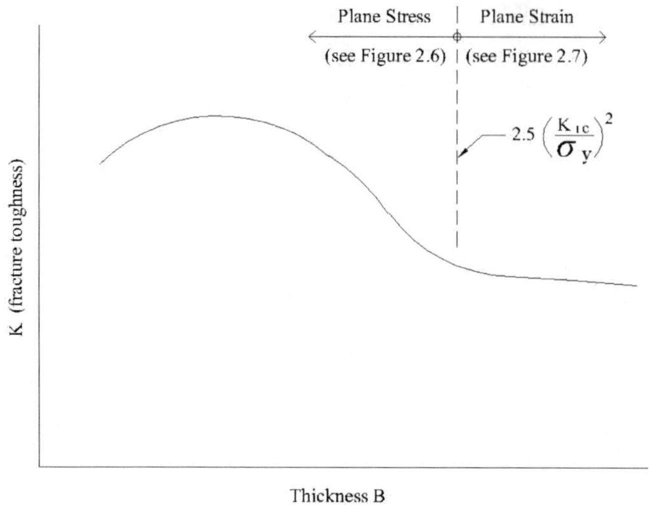

Relationship of Fracture Toughness to Thickness
[McEvily (2001), reproduced with permission of Wiley and Sons, Inc.]

Figure 2.8

often the published value of K_{IC} which represents the lower limit of the fracture toughness of the material based on thickness. Conservative results may occur when β does not conform to the relationship above (i.e. for thin plates).

As discussed above, steel requires dislocation along principal shear planes for ductile behavior. From consideration of the principal stresses shown in Figure 2.9a and Mohr's circle diagram applied to two dimensional stress conditions (see Figure 2.9b), constrained conditions do not allow this to take place. For the two dimensional stress conditions with Stress σ_1 applied, if Stress σ_2 is small (or less than zero), as shown in Figure 2.9b, Mohr's circle is above $\frac{1}{\sqrt{3}} \bullet \sigma_y$ and ductile behavior can occur (assuming σ_3 is also small). As Stress σ_2 increases towards Stress σ_1, Mohr's circle reduces in size and becomes less than $\frac{1}{\sqrt{3}} \bullet \sigma_y$. As a result, shear yielding cannot occur. Constraint in the σ_3 direction further reduces the possibility of ductile behavior. Examples of this condition are the Ingram Barge [see Chapter 1, Section 1.1.6] and the pre-Northridge steel beam to column moment frame connection [see Chapter 1, Section 1.2.11]. This issue was discussed in significant detail by Blodgett (1998) including the more complex consideration of stresses in three directions.

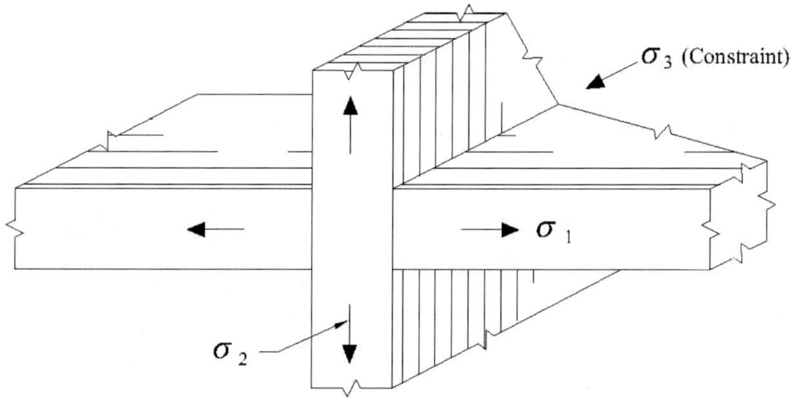

Three-Dimensional Stress Diagram

Figure 2.9a

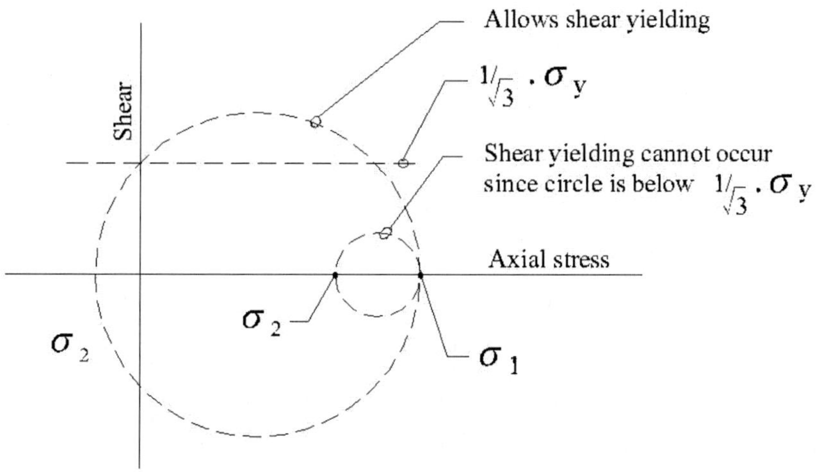

Mohr's Circle Diagram
(two-dimensional)

Figure 2.9b

2.2.4 Effects of Temperature

During World War II, the failures of welded ships ("Liberty" Ships) demonstrated that steel could be very brittle at low temperatures as described in Chapter 1, Section 1.2.2. Constance Tipper, working at the Engineering Department in Cambridge, England, discovered that there is a critical temperature below which steel material is subject to rapid change in ductile to brittle characteristics (see Figure 2.10a). The temperature range at which the significant change occurs is called the nil-ductility temperature. There are several factors influencing the ductile to brittle transition curve including composition, strength level, thickness, and strain rate. The transition temperature is increased with higher levels in steel of carbon, phosphorous, molybdenum and arsenic. The addition of nickel, silicon, manganese and copper decreases the transition temperature.

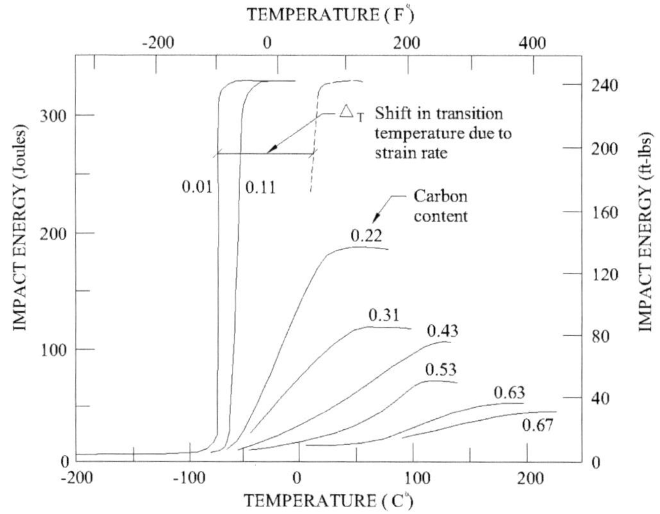

Brittle / Ductile Transition Curves

Figure 2.10a

Based upon extensive studies on service failures and large scale tests conducted by the Batelle Memorial Institute, a Fracture Analysis Diagram (FAD) evolved around 1960 [Pellini (1971), Parts I & II]. The FAD, shown in Figure 2.10b, indicates stress limitation with flaw size and temperature. Below the nil ductility temperature (NDT) the stress limitation is constant. Above the NDT, the stress limitation increases appreciably. Eventually, the curves for the different flaw sizes converge at a crack arresting temperature of 62°C (144°F) above yield stress. This information did not include heavier and/or larger sections where plane strain conditions can prevail.

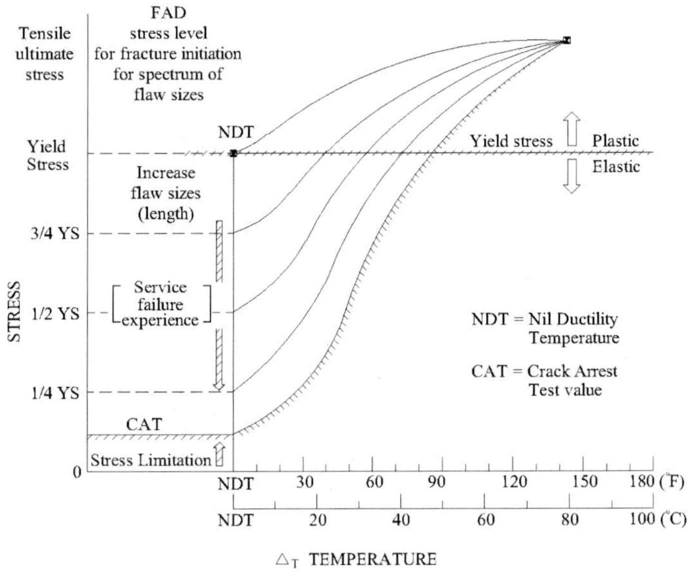

Fracture Analysis Diagram (FAD)

Figure 2.10b

2.2.5 Effects of Strain Rate

Loading rates can be categorized into slow load (strain rate ≈ 10^{-3} per sec.) and dynamic load (strain rate 10 per second).

Fracture Toughness characteristics tend to improve with increasing temperature. However, fracture toughness characteristics decrease with increasing loading rate.

Also, a high strain rate can be such as to significantly affect the ductile to brittle transition temperature. High strain rates tend to shift the transition temperature upwards which can have an adverse effect on performance. According to Barsom and Rolphe (1999), the magnitude of the temperature shift between slow loading and impact loading in steels with various steel yield strengths can be approximated by:

$$T_{shift} = 215 - 1.5\sigma_y \text{, °F}$$

Where:

σ_y = yield stress of the steel in ksi

Thus the lower the yield strength, the greater the temperature shift. For example, for $\sigma_y = 50$ ksi, $T_{shift} = 140°F$ $\left[\sigma_y = 345\,MPa, T_{shift} = 60°C\right]$

According to Barsom and Rolphe (1999), the shift in the transition temperature corresponding to 1 second loading to fracture is approximately 25 percent of the temperature shift determined above for impact loading.

2.2.6 Testing for Fracture

The most common test used in the building industry for estimating steel fracture toughness is the Charpy Vee Notch (CVN) Test which was developed around 1905. The CVN Test is an impact test comprising 10mm (3/8 inch) x 10mm (3/8 inch) x 55mm (2-3/16 inch) long specimens with a notch at midlength transverse to the specimen (see Figure 2.11). The specimen is either cooled or heated to a specified temperature. The specimen is impacted with a striker on the opposite side of the notch usually causing fracture of the specimen. There are formulae that correlate CVN testing with fracture toughness [Barsom and Rolphe (1999), Wallin Correlation given in British Standard (7910)]. These correlations are not precise.

A variety of fracture tests were developed between 1940 and 1950 including the Explosion Crack Starter Test. This test involved a short, brittle weld bead on a plate, which was placed over a circular die and loaded by explosion. Another test, developed in the late 1940's, was the Robertson Crack Arrest Test involving a forced initiation of fracture, which, based upon defined elastic stresses, propagates through a flat plate. By varying temperatures and levels of stress this test was used to establish relationships of fracture with stress, temperature and flaw size as described in Section 2.2.4.

Limitations of the Explosion Crack Starter Test led Pellini, in 1953, to develop the drop weight test at the Naval Research Laboratory, which has been primarily used to determine the nil-ductility temperature (see Section 2.2.4). This involves dropping a weight onto a specimen, which may be welded. Methods were established to determine fracture toughness from the drop weight test. A variation of this test is the Batelle drop-weight tear test using a specimen with a machined notch and which was developed around 1962. The percentage of shear fracture across the failure surface is determined for specimens at different temperatures. The temperature at which the shear fracture surface meets a required percentage is then determined. Drop hammer tests were carried out on 3.66 m (12 feet) long steel beams with welded splices at mid span at Columbia University in the early 1950's by W.J. Krefeld [Grover (1954)]. Beams, tested under low temperature impact, failed by brittle fracture some distance away from the welded splice.

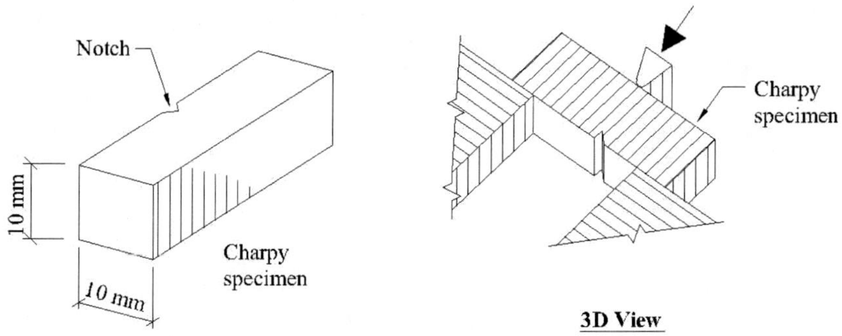

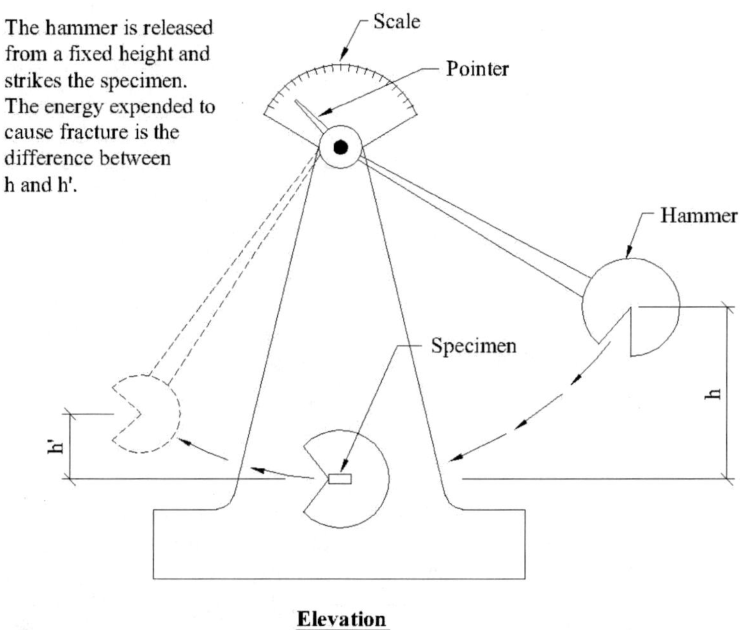

Charpy Vee Notch Test

Figure 2.11

There are other testing methods for estimating fracture toughness. These include the Tipper Test, the Van der Veen Test, the Lehigh Test, the Kinzell Test and the Kammerell Test [Masabuchi (1980)].

The Crack Tip Opening Displacement (CTOD) test, originally known as the Crack Opening Displacement (COD) test was first developed by Wells in the United Kingdom in the 1960s (see Figure 2.12a) but was quite similar to the

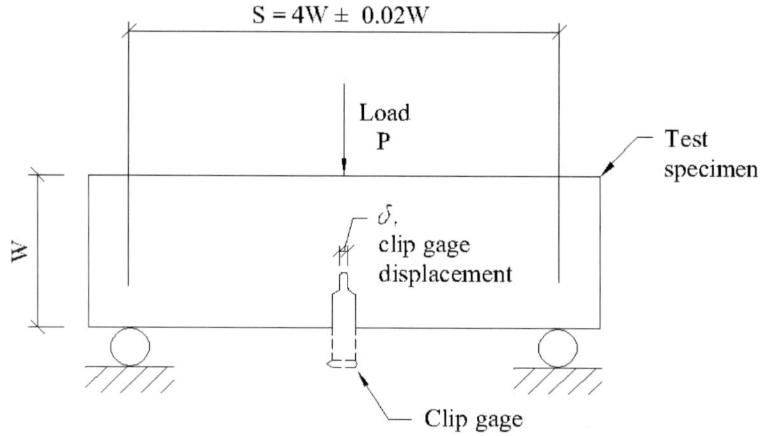

Crack Tip Opening Displacement (CTOD) Test Setup

Figure 2.12a

Robertson Crack Arrest Test mentioned above. It has similarities with the Tipper Test and involves a specimen with a machined notch with loading applied to bend the specimen. The objective of the test is to measure the crack extension to the onset of brittle failure. The load versus crack opening displacement is measured, utilizing a clip gauge, from which maximum allowable crack sizes can be determined applying an appropriate factor of safety. Rice, in the United States, developed a method using "J" integral analysis, applied to the CTOD tests, in order to investigate elastic-plastic conditions. The value for "J" can be determined from the area of the loading and unloading curve (area A shown in Figure 2.12b). Fracture toughness can be determined from the "J" value.

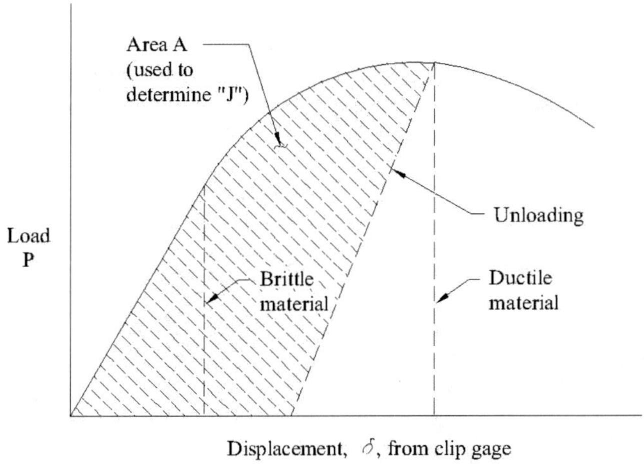

Load/Displacement Relationship for CTOD Test

Figure 2.12b

Approximate relationships for ductile materials may be represented as follows [Dowling (1999)]:

$$\delta \approx \frac{K^2}{E\sigma_y} \approx \frac{J}{\sigma_y}$$

Where:

δ = CTOD
K = Fracture toughness
σ_y = yield strength

Although there are different standards for the CTOD Test in the USA, Europe and the United Kingdom, the tests in these standards are very similar.

In more recent years, the crack tip opening angle (CTOA) geometric parameter, based upon the CTOD test, has been used to assess fully plastic fracture [(Darcis et al (2007)]. The comparison between CTOD and CTOA is shown in Figure 2.12c. Measurements of the CTOA are determined from digital images. The criteria for CTOA is based upon the following relationship:

$$CTOA_{\max} < CTOA_c$$

Where:

$CTOA_{max}$ = measure of the maximum crack driving force calculated from the geometry, material properties and loading conditions

$CTOA_c$ = resistance of the material to crack growth

Darcis et al describe the application of the CTOA method for measuring pipe weld toughness.

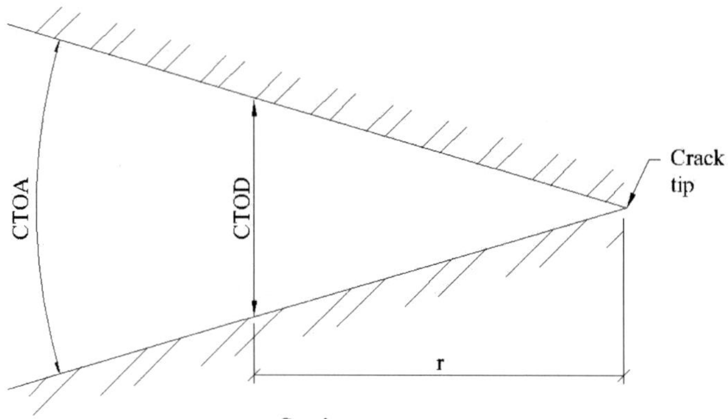

Section

Comparison of CTOA with CTOD Measurement

Figure 2.12c

2.2.7 Temperature Increase Due to Dynamic Behavior and Yielding

During yielding and dynamic loading, plastic work is transferred to heat. The amount of temperature increase is somewhat dependent upon the sulphur content and distance from a notch (defect) where plastic work is occurring according to Ishikawa et al (1998). Lower temperature increase occurs with higher sulphur content and greater distance from the notch.

As mentioned in Section 2.2.5, the ductile to brittle transition temperature increases with increase in strain rate. Furthermore, as discussed in Chapter 3, Section 3.12, increase in strain rate and temperature increases the yield strength of steel.

Assessing the combination of the phenomena mentioned above is highly complex. Attempts at taking these combined factors into account, also including for residual stresses, has been made by Shimanuki and Hagiwara (1998) on damaged steel framed structures following the 1994 Northridge and 1995 Hyogoken – Nanbu (Kobe) earthquakes.

2.2.8 Applicability of Linear Elastic Fracture Mechanics (LEFM)

LEFM is based on the assumption that the plastic stress field at the crack tip is sufficiently small that principles of linear elasticity will apply and are dominant beyond the plastic zone (see Section 2.2.2, Figures 2.2d and Figure 2.3). LEFM is applicable if the following relation holds true:

$$a,(b-a),h \geq \frac{4}{\pi}\left(\frac{K}{\sigma_y}\right)^2$$

Where:

h = distance from the crack (horizontal) centerline to the nearest edge, in the direction of loading

Satisfaction of the above criteria will, according to LEFM, preclude the formation of a through-member plastic zone.

If the above criteria are not satisfied, the crack tip stress field will tend to be dominated by a large plastic zone that may ultimately extend to a boundary. In this case, LEFM may underestimate the severity of the crack.

2.3 FATIGUE

2.3.1 General Discussion

Fatigue is a form of fracture failure that occurs when a material is subjected to repeated or cyclic tensile loading. Structural components subjected to repetitive loading may fail at stresses well below the tensile yield stress of the material. Evidence of fatigue failures was found in the early 19th Century (circa 1829) including tests by W.A.J. Albert in Germany on mine hoist chains. Further work and discussion was carried out when train accidents occurred due to fracture of railroad car axles. An extensive study on fatigue was carried out by August Wöhler in Germany on axles (made from iron, steel and other materials) subjected to tensile, bending and torsional load fluctuations during the period between 1852 to 1870 [Ref. Barsom and Rolphe (1999), Dowling (1999), McElivy (2001), Munse (1964) and Lancaster (1992)]. The accumulated damage due to load fluctuations results in initiation of cracks and repeated fluctuations can result in the propagation of cracks leading to failure. A particular example is at the toe of fillet weld, or where a defect occurs at a backup bar of a groove weld. The resulting defect tends to act as a point of initiation and when subjected to cyclic loading, can lead to crack propagation.

In general there are three "phases" that occur to produce a fatigue failure. The first phase of *crack initiation* occurs from cyclic loading of material from the virgin condition to the formation of a macro-crack. The *crack propagation* phase is characterized by the stable growth of the crack from the crack initiation phase. The final phase of fatigue failure is fracture, characterized by rapid, unstable crack growth.

A number of types of fatigue failure are known to exist. The types of fatigue that most readily affect engineered structures are briefly mentioned in the next paragraph. The discussion is not exhaustive.

Mechanical fatigue occurs solely due to fluctuations in stress. *Thermal fatigue* occurs when a component is subjected to repeated heating and cooling. In this mode, cyclic stresses are caused by stresses that occur as a result of differential thermal expansion/contraction. *Corrosion fatigue* occurs in components subjected to cyclic loading and aggressive chemical environments. The effects of corrosion are such as to result in the pitting of the metal surfaces which then causes stress raisers. The stress raisers cause a reduction in the fatigue life of the steel. This can result in the fatigue life being somewhat independent of the ultimate strength of steel. Adequate protective measures (e.g. painting under mild conditions, cathodic protection under more severe conditions) must be provided to eliminate pitting. Please refer to the references cited in this document for additional discussion on these topics. The discussion of fatigue in this document will be confined to mechanical fatigue. Henceforth, the term *fatigue* will necessarily imply the phenomenon of mechanical fatigue.

To summarize fatigue again, it may be represented in three phases as follows:

(1) Crack initiation
(2) Crack propagation
(3) Unstable crack growth leading to final failure

It should be noted that pre-existing defects could eliminate the crack initiation stage and thereby decrease the total fatigue life.

The three steps, described above, can be represented as follows:

$$N_f = N_i + N_p$$

Where:

N_f = number of cycles to failure
N_i = number of cycles for crack initiation
N_p = number of cycles for crack propagation (to failure)

Fatigue may be further characterized by the number of load repetitions (or cycles) that a component is subjected to prior to failure. When a component is expected to survive a relatively large number of cycles the concept of *high-*

cycle fatigue is often used to address this situation. When the number of cycles is not large the concept of *low-cycle fatigue* may be used to address the low-cycle condition.

The primary difference between the high-cycle and low-cycle fatigue is the stress regime in which each is loaded. A component subjected to high-cycle repetitive loading typically operates within the linear elastic (and monotonically loaded) regime of the stress-strain relation for the material (see Figures 2.13 and 2.14a). The stress associated with the low-cycle fatigue regime may be characterized with excursions into the inelastic regime (and a stress-strain relationship characterized by a hysteretic "loop") of the stress-strain relation (see Figure 2.14b). In general, a *stress-based* approach is used to consider component life in the high-cycle regime while a *strain-based* approach is used to assess a low-cycle fatigue life. It should be noted that the strain-based approach is applicable to both high and low-cycle fatigue situations. The same is not true for the stress-based approach. However, the stress-based approach is often more readily applied to high-cycle situations in practice. Methods for assessing both high and low cycle fatigue issues will be presented in this document.

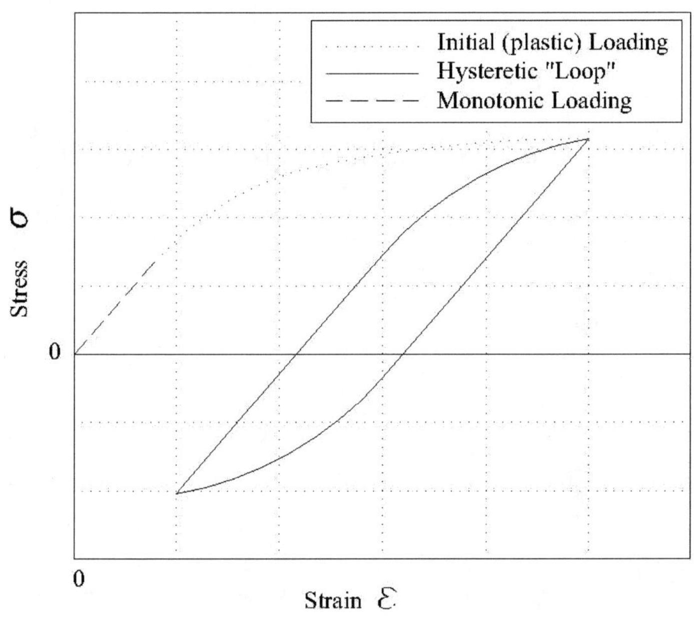

Stress/Strain Relationship (associated with Cyclic Loading)

Figure 2.13

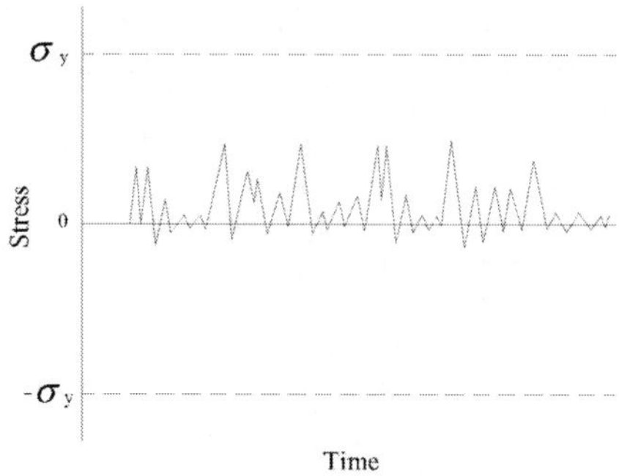

High Cycle Fatigue

Figure 2.14a

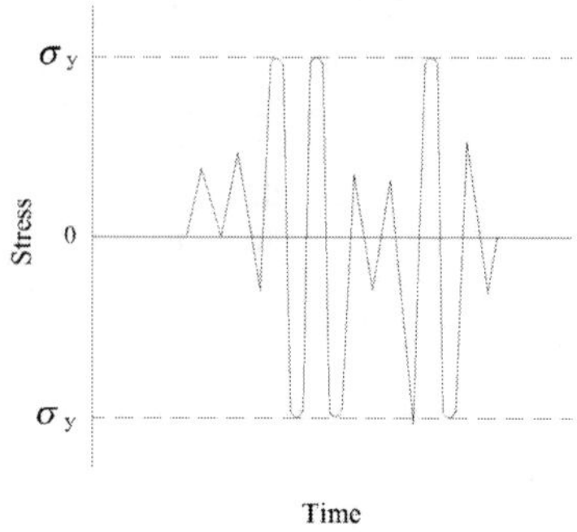

Low Cycle Fatigue

Figure 2.14b

If a crack or crack-like flaw is already present in a component subjected to cyclic loading, the concept of *fatigue crack growth* must be explored. The rate at which the crack will grow as well as the number of cycles at which the crack length becomes critical is of concern. A method for approaching this concern, based on LEFM, will be presented later in this document.

2.3.2 High-Cycle Fatigue

As was mentioned previously, high cycle fatigue is generally considered to occur when a component is subjected to a large number of load cycles. A stress-based approach is typically used to assess fatigue life in the high cycle regime. To apply this approach to fatigue it is necessary to characterize the nature of the loading (or stressing) of the component with the following definitions:

Stress Range: $\Delta\sigma = \sigma_{\max} - \sigma_{\min}$

Mean Stress: $\sigma_m = \dfrac{\sigma_{\max} + \sigma_{\min}}{2}$

Stress Amplitude: $\sigma_a = \dfrac{\Delta\sigma}{2}$

Where:

$\sigma_{\max}$ = maximum stress that occurs during the load cycle

$\sigma_{\min}$ = minimum stress that occurs during the load cycle

Please refer to Figure 2.15.

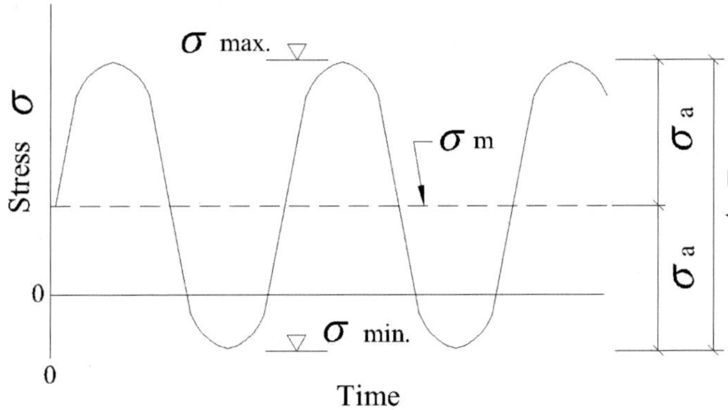

Constant Amplitude Cycling

Figure 2.15

Application of the stress-based approach to fatigue relies upon having appropriate material information. Specific to fatigue concerns is the concept of a Stress-Versus-Life curve also known as an S-N_f curve. The S-N_f curve is constructed experimentally by testing material specimens under cyclic loading at various levels of stress. The result is a plot of stress (S) versus the number of cycles to failure (N_f). The relationship between a particular stress and cycles to failure (or vice versa) can be readily determined from an S-N_f Curve.

Figure 2.16 indicates the typical S-N_f curve for low-alloy steels. It should be noted that for steel, there is a distinct stress level, known as the "Fatigue Limit", below which fatigue failure does not occur.

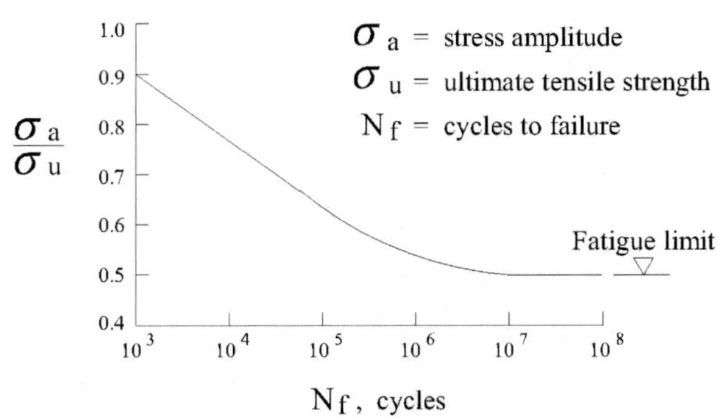

Typical S-N Behavior for Low Alloy Steels
[McEvily (2001), reproduced with permission of Wiley and Sons, Inc.]

Figure 2.16

With the appropriate material properties, the S-N_f, curve for a material can be described with the following equation:

$$\sigma_a = \sigma'_f \left(2N_f\right)^b$$

Where:

N_f = number of cycles to failure

σ'_f, b = material properties (σ'_f may be approximated with the true fracture strength of the material)

Unless otherwise stated, an S-N_f curve typically represents the relationship between S and N_f for a condition of completely reversed stressing in which:

$$\sigma_{\min} = -\sigma_{\max} \text{ and } \sigma_m = 0$$

In practice, it is often the case that stressing is not completely reversed. Moreover, it can be demonstrated experimentally that, for a net mean tensile stress, the fatigue life can be much less than predicted for completely reversed stress cycles of the same amplitude.

One method of addressing cyclic loading that has a mean tensile stress is to convert the stress amplitude, σ_a, to an equivalent completely reversed stress amplitude, σ_{ar}. The value of σ_{ar} can be used with an appropriate S-N_f curve to determine the fatigue life of the material for the load cycles under consideration. An equation for this conversion that is often used with ductile metals and which was first proposed by J. Morrow (circa 1968), is [Dowling (1999)]:

$$\sigma_{ar} = \frac{\sigma_a}{1 - \dfrac{\sigma_m}{\sigma'_f}}$$

The previous discussion provides provisions that may be applied to stresses from uniaxial or bending load conditions. In practice, it is conceivable that a cyclic loading may produce a multiaxial stress condition. Such a condition may arise, for example, in the analysis of steel connections. The multiaxial state of stress may be determined by a finite element analysis. One method of addressing a cyclic multiaxial state of stress is to resolve the multiaxial state into a resultant tensile stress. A general form, developed from the state of axial and shear stresses in three planes, which may be used to calculate the net tensile stress amplitude for a state of stress is:

$$\sigma_a^{ma} = \frac{1}{\sqrt{2}}\sqrt{(\sigma_{1a} - \sigma_{2a})^2 + (\sigma_{2a} - \sigma_{3a})^2 + (\sigma_{3a} - \sigma_{1a})^2 + 6(\tau_{12a}^2 + \tau_{23a}^2 + \tau_{31a}^2)}$$

Where:

σ_{ia} = normal stress amplitude in the ith principle direction

τ_{ija} = shear stress amplitude in the ij plane

The value of σ_a^{ma} and an appropriate S-N curve may be used to determine the allowable number of cycles to which a component may be subjected. In the presence of non-zero mean stresses the following relation may be used to determine an effective mean stress:

$$\sigma_m^{ma} = \sigma_{1m} + \sigma_{2m} + \sigma_{3m}$$

Where:

σ_{im} = mean stress in the ith principle direction

σ_a^{ma} and σ_m^{ma} can be used in the manner described above to determine the equivalent completely reversed stress amplitude. Explicitly described as:

$$\sigma_{ar} = \frac{\sigma_a^{ma}}{1 - \dfrac{\sigma_m^{ma}}{\sigma'_f}}$$

The effect of a notch in a component made of ductile material can be considered with the following relation:

$$\sigma_{ar} = \frac{k_f S_a}{1 - \dfrac{S_m}{\sigma'_f}}$$

Where:

k_f = fatigue stress concentration factor

S_a = stress amplitude on the gross area of the component

S_m = gross mean stress from the cyclic loading

It should be noted that the fatigue stress concentration factor, k_f, is generally not equal to the stress concentration factor associated with the static load case. Tables and graphs that define the fatigue stress concentration factor for various load conditions may be found in references at the end of this section.

In practice, it may be necessary to assess the fatigue life of a component subject to variable amplitude loading. The most common way to address this issue is with the concept of a fatigue life fraction:

$$\frac{n_i}{N_{fi}} = \text{Fatigue Life Fraction}$$

Where:

n_i = number of load cycles incurred by the component at a stress amplitude i

N_{fi} = number of load cycles that may be incurred by the component at the stress level i or the "fatigue life"

This ratio represents the number of load cycles incurred to the number of allowable load cycles for a particular stress amplitude. Stated differently, this ratio represents the total amount of fatigue life that has been "used up". Theoretically, the fatigue life has ended when the ratio equals unity.

When a component is subjected to varying load amplitudes, the sum of the fatigue life fractions from each of the load amplitudes may be added directly to assess the total fatigue life fraction. This concept is commonly referred to as the Palmgren-Miner Rule and may be stated as follows:

$$\sum_j \frac{n_j}{N_{fj}} = \text{Total Fatigue Life fraction}$$

Where:

n_j = number of load cycles of amplitude j incurred by the component

N_{fj} = fatigue life at amplitude j

As before, fatigue life will theoretically end when the above sum equals unity.

Therefore, for safe design $\sum_j \frac{n_j}{N_{fj}} < 1$ with an appropriate factor of safety applied.

An excellent reference on fatigue, with an emphasis on high cycle fatigue pertaining to steel bridges, is given by Fisher et al (1998).

2.3.3 Low Cycle Fatigue

Low cycle fatigue can be characterized with load cycles that cause stresses and strains in the inelastic regime of a stress-strain relation as previously discussed in Section 2.3.1 and shown in Figure 2.14b. Figure 2.17a shows a

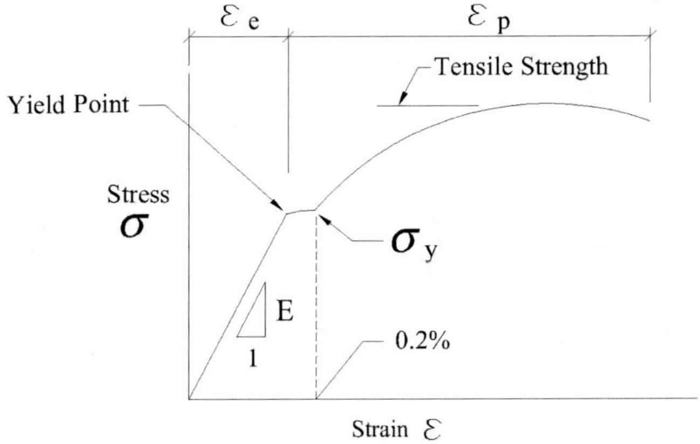

Stress/Strain Relationship

[Reproduced from Dowling, Norman E., Mechanical Behavior of Materials, 3rd Edition (2007), reprinted with permission of Pearson Education Inc., Upper Saddle River, NJ]

Figure 2.17a

typical stress-strain relationship for steel. Beyond yielding, strain can be represented by elastic and inelastic components.

$$\varepsilon = \varepsilon_e + \varepsilon_p$$

Where:

ε = total strain
ε_e = elastic strain
ε_p = plastic strain

The elastic strain can be described with the following constitutive relation:

$$\varepsilon_e = \frac{\sigma}{E}$$

The plastic strain may described with the exponential relation:

$$\varepsilon_p = \left(\frac{\sigma}{H}\right)^{\frac{1}{n}}$$

Where:

σ = stress
E = Modulus of Elasticity
H, n = material constants

Figure 2.17b shows the relationship in logarithmic coordinates.

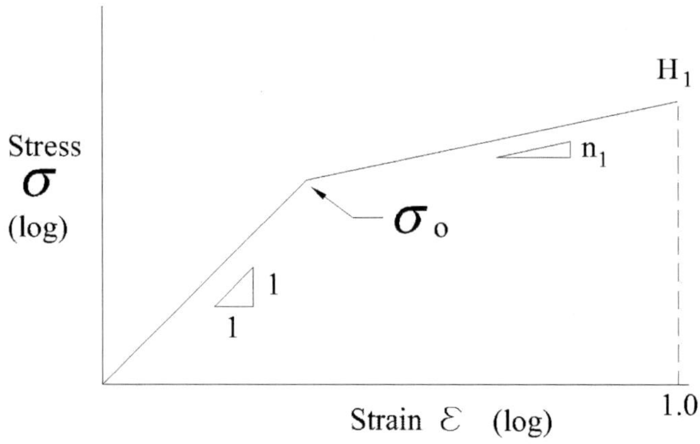

Stress/Strain Relationship
[Reproduced from Dowling, Norman E., Mechanical Behavior of Materials, 3rd Edition (2007), reprinted with permission of Pearson Education Inc., Upper Saddle River, NJ]

Figure 2.17b

The material constants, H and n, are derived from test data for the plastic regime of a stress-strain plot. Steel displays strain hardening characteristics with a positive n value.

Combining the above equations results in a form of the Ramberg-Osgood Relationship (proposed in a report by them in 1943).

$$\varepsilon = \frac{\sigma}{E} + \left(\frac{\sigma}{H}\right)^{\frac{1}{n}}$$

This equation represents a smooth curve that describes the total strain of a material into the plastic regime (see Figure 2.17c). Such an equation can be used to construct a model combining elastic and plastic behavior for a given load history. Combinations of two or more behaviors are called rheological models.

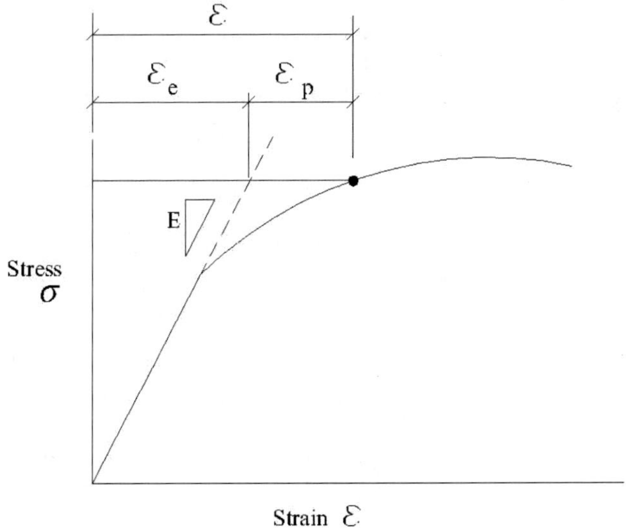

Ramberg-Osgood Relationship

Figure 2.17c

During cyclic loading, with yielding and unloading occurring, the stress-strain path differs from the preceding cycle. A cyclic stress-strain rheological model can also be developed. In this case, the Ramberg-Osgood relationship can be used and takes on the special form of:

$$\varepsilon_a = \frac{\sigma_a}{E} + \left(\frac{\sigma_a}{H'}\right)^{\frac{1}{n'}}$$

Where:

ε_a = the strain amplitude

σ_a = the stress amplitude

E = the Modulus of Elasticity

H', n' = material constants for cyclic loading

It should be noted that, for steel, when first deformed in tension, then deformed in compression, the compressive strength is lower than if only tested in compression (see Figure 2.18). This property, known as Bauschinger effect (after a German Engineer who discovered the phenomena in the 1880's), also occurs when compression is first applied and the load is reversed into tension.

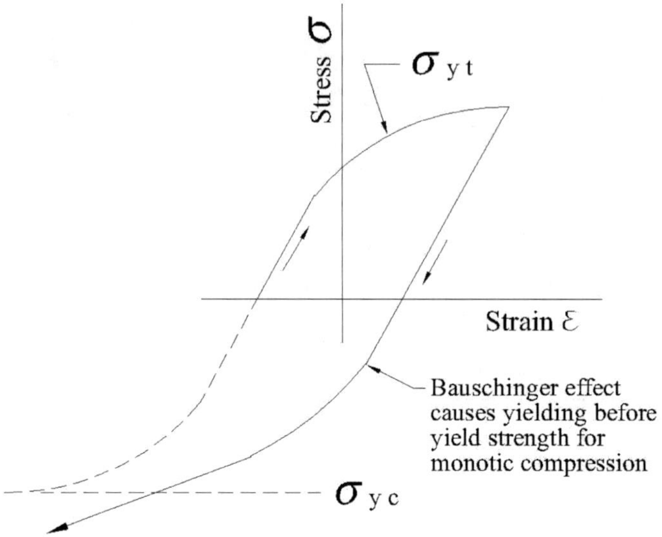

Cyclic Stress/Strain Curve: Bauschinger Effect
[Reproduced from Dowling, Norman E., Mechanical Behavior of Materials, 3rd Edition (2007), reprinted with permission of Pearson Education Inc., Upper Saddle River, NJ]

Figure 2.18

Like stress-life data, a procedure exists for developing strain-life data. The testing procedure consists of a completely reversed amplitude cycling between constant strain limits. The data from such tests are plotted and used to define an empirical relationship between plastic strain and number of cycles. This relation is:

$$\varepsilon_{pa} = \varepsilon_f^{'} \left(2N_f\right)^c$$

Where:

ε_{pa} = cyclic strain amplitude

ε'_f = intercept constant for the plastic strain at zero cycles (see Figure 2.19)

N_f = number of cycles to failure

c = material property

Substituting this equation and the stress-life relationship from the High-Cycle Fatigue section yields the following relation:

$$\varepsilon_a = \frac{\sigma'_f}{E}\left(2N_f\right)^b + \varepsilon'_f\left(2N_f\right)^c$$

This equation is referred to as the Coffin-Manson Relationship and is represented in Figure 2.19. The above expression gives a relationship between number-of-cycles and total strain amplitude. Where the lines for the elastic and plastic relationships intersect is known as the transition fatigue life. Thus, the fatigue life to the left of the intersection indicates low cycle fatigue, whereas to the right of the intersection represents high cycle fatigue.

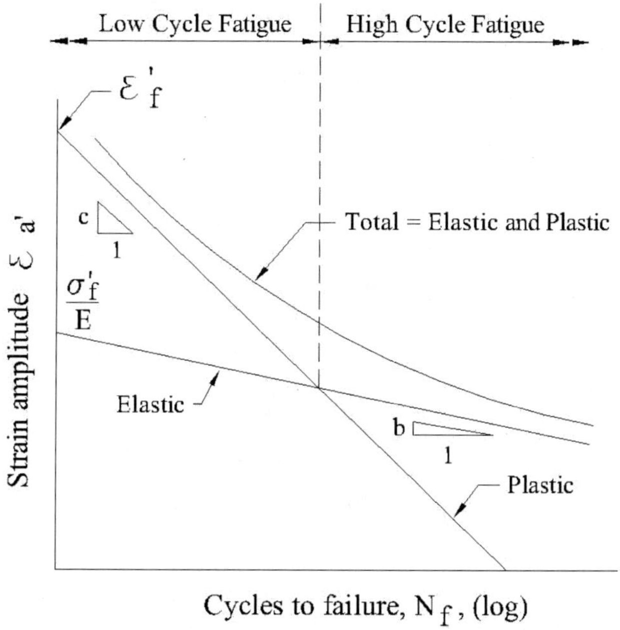

Elastic, Plastic and Total Strain versus Number of Cycles: Coffin-Manson Relationship

[Reproduced from Dowling, Norman E., Mechanical Behavior of Materials, 3rd Edition (2007), reprinted with permission of Pearson Education Inc., Upper Saddle River, NJ]

Figure 2.19

A ductile material tends to have a high strain capacity but low strength. Thus failures due to a low number of cycles occur at high strain. The lower strength characteristic of a ductile material results in low elastic strain at failure due to a high number of cycles. This results in a steep curve for the relationship between strain amplitude and number of cycles for a ductile material. However, for a brittle material, which has high strength but low strain capacity, the curve is shallow. These relationships, along with a tough material, representing an intermediate case, are indicated in Figure 2.20. It is interesting to note that, based upon testing, at approximately 1000 cycles the stress-strain curves for a wide variety of materials all pass through a point with an accumulated strain of about 0.01 [Dowling (1999)].

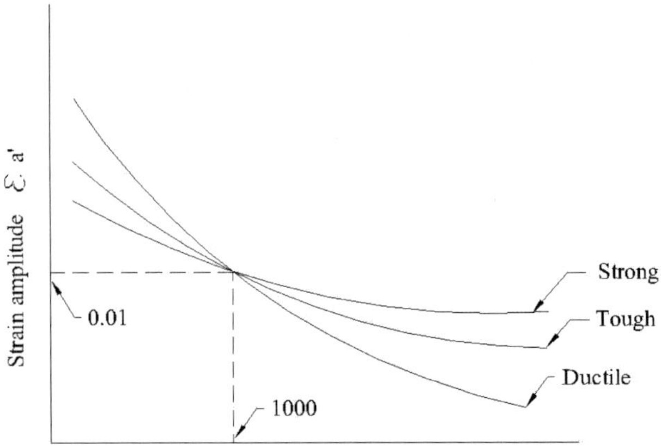

Trends in Strain-Life Curves from Strong, Tough and Ductile Metals
[Reproduced from Dowling, Norman E., Mechanical Behavior of Materials, 3rd Edition (2007), reprinted with permission of Pearson Education Inc., Upper Saddle River, NJ]

Figure 2.20

The fatigue life will be modified if a mean stress other than zero stress occurs. If strain amplitudes are not large, the material remains mostly elastic and much of the mean stress remains. Mean strain would not significantly affect fatigue life unless it is quite large.

The following relationship has been derived to account for the effects of mean stress:

$$\varepsilon_a = \frac{\sigma'_f}{E}\left(1 - \frac{\sigma_m}{\sigma'_f}\right)(2N_f)^b + \varepsilon'_f\left(1 - \frac{\sigma_m}{\sigma'_f}\right)^{\frac{c}{b}}(2N_f)^c$$

2.3.4 Fatigue Crack Growth

As was stated in section 2.2, the presence of a crack (or defect) in a component can greatly reduce the strength of the component due to the potential for non-ductile failure. The presence of a crack in a component subjected to cyclic loading is of concern. A crack subjected to cyclic loading can grow as additional cycles are incurred, leading to fast unstable crack growth and ultimately fracture. The concept of a crack growing under cyclic loading is known as *fatigue crack growth.* Recalling the relationship from Section 2.2.2:

$$K = YS\sqrt{\pi a}$$

The '*Y*' term is a function of the ratio of *a/b*, where *a* is the crack length and *b* is the component dimension parallel to the crack. The rate of fatigue crack growth is controlled by K, the stress intensity factor. Considering the dependence of *K* on *Y* and *a*, it is apparent that the rate of crack growth increases as the crack length, *a*, increases. Recalling the relationship from Section 2.2.2:

$$S_a = \frac{K_c}{\sqrt{\pi a}}$$

Where:

K_c = fracture toughness of the material

It is apparent that the allowable stress, S_a, will decrease as a increases. Considering a cyclic load, failure will occur when $\sigma_{\max}$ and S_a are equivalent.

One way to characterize the manner in which a crack will grow under cyclic loading is to introduce the following ratio:

$$\frac{da}{dN}$$

Where:

da = change in crack length
dN = number of cycles corresponding to the change in crack length

This relationship is referred to as the *cyclic crack growth rate (Dowling (1999)*. A further relationship exists between the cyclic crack growth rate and the change in the stress intensity factor.

$$\frac{da}{dN} = C(\Delta K)^m$$

Where:

ΔK = change in stress intensity factor
C, m = fitting constants from experimental data from fatigue crack growth rates of a particular material

The relationship of crack length to number of cycles and the crack growth rate along with the change in the stress intensity factor is shown in Figure 2.21. It can be seen that initially the curve is approximately constant until the crack length is such as to cause a greater rate of increase leading to failure. The point on the curve where this distinct change occurs is called the *fatigue crack growth threshold.*

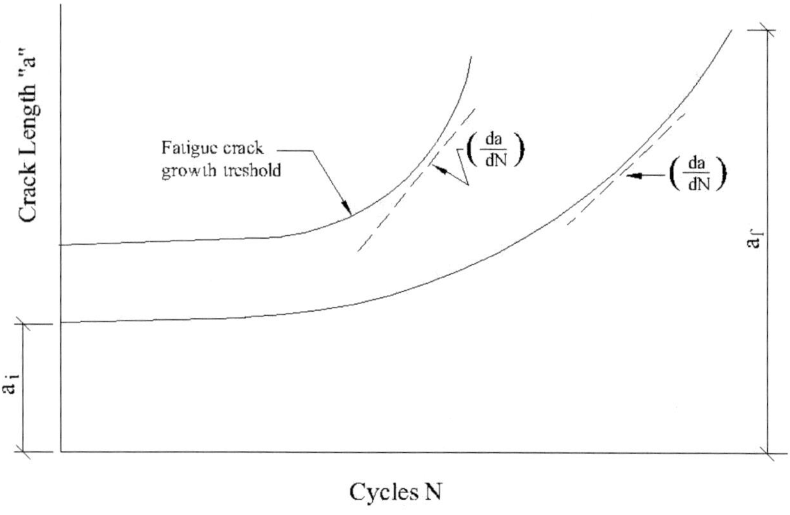

Crack Length versus Number of Cycles at Stress Levels

Figure 2.21

The curve for number of cycles to failure, as shown in Figure 2.21, may be represented by the following expression derived from the equation above:

$$N_f = \frac{a_f^{1-\frac{m}{2}} - a_i^{1-\frac{m}{2}}}{C\left(Y \Delta S \sqrt{\pi}\right)^m \left(1 - \frac{m}{2}\right)}$$

Where:

N_f = number of load cycles required to "grow" the crack from length a_i to a_f

a_i = initial crack length

a_f = final crack length (e.g. crack length at failure)

ΔS = stress amplitude on the gross cross section of the component

The above expression is for loading cycles that cycle from zero to $\sigma_{\max}$. The empirical constants for the above expression are typically derived from data for a load cycle of this nature.

Fatigue-crack propagation in steel can also be represented as falling into three regions as indicated in Figure 2.22. Figure 2.22 shows the logarithmic relationship between crack growth rate, da/dN, and Stress Intensity Factor typical for steel. The three regions are as follows:

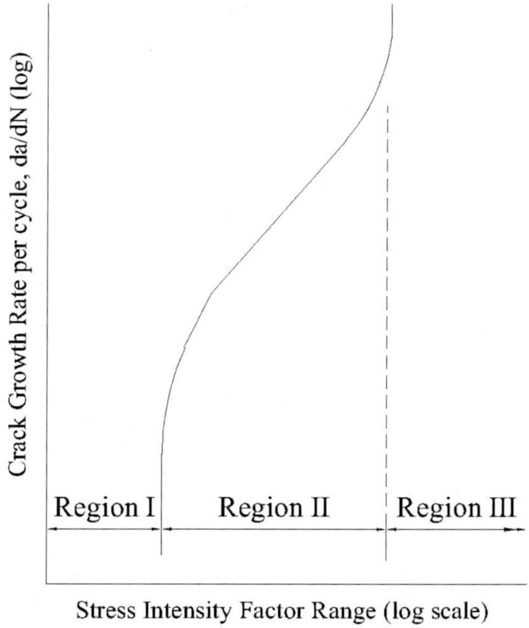

Relationship of Crack Growth to Stress Intensity Factor
[Reproduced with permission from Facture and Fatigue Control in Structures by Barsom and Rolphe, 3rd Edition, copyright, ASTM International]

Figure 2.22

Region I: cracks do not propagate.

Region II: $\frac{da}{dN} = C(\Delta K)^m$ (the equation previously indicated)

Region III: Higher than Region II

The above approach, incorporating testing, can be used to assess tolerable flaw sizes based upon predicted stress life cycles and material fracture toughness. This can be applied to steel components, connection details including welds and heat affected zones.

Summarizing, in practical terms, the fatigue life of a steel structure is subjected to the following:

1. The number of cycles applied.
2. The type of detail that may cause stress and/or strain concentrations.
3. The stress/strain range applied at the detail.

2.4 ANALYSIS OF FAILURES

In order to analyze failures, essentially an investigation similar to a forensic investigation needs to be carried out. The investigation usually includes collection of samples and background data, visual examination and photography. Hardness and chemical analysis may also be necessary. Examination of samples can be carried out at a macroscopic level. Electron microscopy and/or energy dispersive spectroscopy may be used to examine fracture surfaces. These procedures, along with analysis using fracture mechanics, and, if necessary, simulated testing of service conditions, can greatly assist in identifying the causes and the location at which the failure initiated.

An excellent overview of the analysis of failures is given in Patel (2008).

RECOMMENDED READING – CHAPTER 2

Barsom, J.M. and Rolphe, S.T., (1999); Fracture and Fatigue Control in Structures, 3rd Edition ASTM.

Dowling, Norman E., (1999); Mechanical Behavior of Materials-Engineering Methods for Deformation, Fracture and Fatigue, 2nd ed., Prentice Hall.

Fisher, J.W., Kulak, G.L. and Smith, I.F.C., (1998); A Fatigue Primer for Structural Engineers; National Steel Bridge Alliance.

Fisher, J.W., Kulak, G.L., Smith, I.F.C., (1997); A Fatigue Primer for Structural Engineers; Advanced Technology for Large Structural Systems (ATSS) Report No. 97-11, October.

Jones, D.A., (1996); Principals and Prevention of Corrosion, 2nd Edition, Pub. Prentice Hall, Upper Saddle River, NJ 07458.

Lancaster, J., (1992); Handbook of Structural Welding; McGraw Hill.

Lancaster, J; Handbook of Structural Welding, Woodhead Publishing Ltd., 1997 (UK).

Masubuchi, K., (1980); Analysis of Welded Structures, MIT, Pergamon Press.

McEvily, A.J., (2001); Metal Failures; Mechanisms Analysis, Prevention, Wiley-Interscience.

Steel Designer's Manual, 5th Edition, 2000, The Steel Construction Institute, Pub. Blackwell Science

CHAPTER 3

STEEL MATERIAL

3.1 METALLURGY OF STEEL

Before describing the history of steel and steel production, it is appropriate to give a brief overview on the metallurgy of steel to give a better understanding of the material and terminology used.

The basic difference between steel and pure iron is that steel contains a certain amount of carbon which reduces ductility and toughness but increases strength. Carbon also increases the susceptibility to hardening when steel is rapidly cooled from elevated temperatures.

The combination of two elements associated with temperature changes can be represented by a Phase Transformation diagram which is important in understanding micro structure properties. The Iron-carbon Phase Transformation Diagram is shown in Figure 3.1. As the iron carbon combination reduces in temperature, it changes from liquid to a partial liquid and solid phase to eventually a solid state. The temperature at which this changes is known as the Peritectic Point. At carbon content levels of 0.8% and 4.3%, the mix changes directly from liquid to solid. These are termed eutectoid points.

The crystal structures (or unit cell), which is how the atoms (or ions) are bonded together, can be two types for the iron carbon combination. These are the body centered cubic (BCC) and face centered cubic (FCC) structures. The BCC structure has ions in each of the four corners and one in the center of the cube. The FCC structure has ions in each of the four corners and one in the center of each face (see Figure 3.2).

There are essentially four changes in iron which are known as alpha (α), beta (β), gamma (γ) and delta (δ). For pure iron, α iron occurs in the solid phase. The α to β change occurs at about 723°C (1334°F) and the β to γ change occurs at about 910°C (1670°F). At a temperature of about 1400°C (2550°F), the iron then changes to δ iron. The α, β and δ forms have the

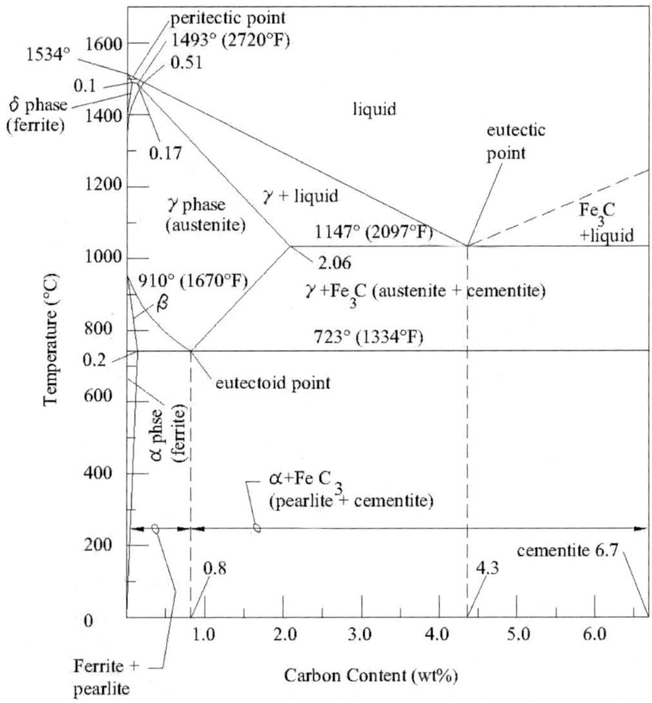

Iron – Carbon Phase Transformation Diagram

Figure 3.1

same body centered cubic structure (BCC). However, γ-form has a face centered cubic structure (FCC). There is a marked contraction when the crystal structure changes from BCC to FCC. Thus, a marked contraction, accompanied with recrystallization followed by grain growth, occurs at the β-iron to γ-iron change. The structure then expands at the γ-iron to δ-iron change.

The purest form of iron, built up from a number of crystals of the same composition, (α-iron) is given the name **ferrite**. Iron and carbon, as a compound (Fe_3C), is represented by the term **cementite**. Ferrite and cementite both tend to be weak. When ferrite and cementite are combined to form a eutectoid mixture, containing 0.8% carbon, it is known as **pearlite**.

Space Latices

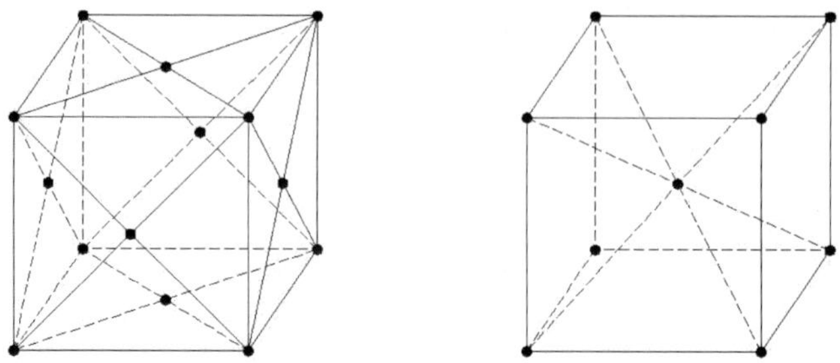

Face Centered Cubic (FCC) Body Centered Cubic (BCC)

Atomic arrangements

Figure 3.2

Steel with about 0.17% carbon contains about 80% ferrite and 20% pearlite whereas steel with about 0.87% carbon contains 100% pearlite. Steel comprising 100% pearlite provides the highest strength. Any further increase in carbon content, above 0.87%, results in free cementite at the grain boundaries or as needles. This free cementite tends to increase hardness but reduces strength. At high temperatures, when the crystal structure is face centered cubic, up to 1.7% carbon can be absorbed with the δ iron to form a solid solution called **austenite**.

When steel is allowed to cool relatively slowly, ferrite and cementite are ejected from the austenite and ferrite starts to form a network structure at the grain boundaries with pearlite occupying the center.

With a faster cooling rate, complete separation of ferrite from the large austenite grains is not possible. A mesh like arrangement known as a Widmanstätten Structure occurs, with ferrite located along octahedral (eight sided) planes. Where steels have a carbon content in excess of 0.9%, cementite can also form Widmanstätten structures. Since the high strength pearlite becomes isolated by the weak ferrite and cementite, cracks can be readily propagated. Thus steels with these type of structures tend to have low impact and low percentage elongation properties.

If the cooling rate is very rapid, as can occur during quenching of the steel with oil (see later), diffusion from austenite to a body centered cubic structure is incomplete and a proportion of austenite is retained in the steel. This structure, which tends to be needlelike, is called **martensite**. High internal stresses occur with a significant density of dislocation. This can cause an expansion of 1 to 3%. The steel properties are very hard with high strength but brittle and with low ductility.

A structure similar to martensite, but less needlelike and occurring in steels containing alloys, is known as **bainite**.

Metallurgical terms may be summarized as follows:

Austenite

Austenite is a face centered cubic structure known as gamma iron. It is a solid solution with 1.7% carbon stable only at high temperatures.

Bainite

Bainite forms when quenching alloy steels. It consists of an aggregate of ferrite and carbide. It has a needlelike structure less pronounced than Martensite (see later).

Cementite

Cementite is a crystalline compound of iron and carbon (Fe_3 C) and is essentially brittle and hard.

Ferrite

Ferrite is essentially pure iron known as alpha iron in the form of a body centered cubic structure.

Martensite

Martensite is formed from quenching and has a fine needlelike structure, and is very hard and brittle.

Pearlite

Pearlite is a mixture of Ferrite and Cementite. It occurs from the transformation of austenite on slow cooling.

3.2 BRIEF HISTORY OF STEEL

3.2.1 Cast Iron

Cast Iron was first developed as early as 200 BC. The basic ingredients of iron are iron ore combined with other iron bearing materials, limestone and coke. The ore comprises iron oxides or carbonate associated with impurities derived from earth that it is removed from. Limestone tends to act as a cleaning agent reacting with the impurities and coke is used as the primary source of carbon. Ore, limestone and coke, in combination as a material, is known as Pig Iron. Cast Iron contains between 2 and 4% carbon along with a significant amount of silicon and other impurities. Consequently, it has low tensile strength and can be brittle. During the 14th century, the first masonry furnace was made in Europe. This enabled significant quantities of iron to be used for castings. During the latter part of the 18th Century, cast iron was used in a number of structures in Great Britain. A particular example is the Severn Bridge, Coolbridge, Wales 1773. Cast iron, which has a high carbon content (more than 1.5%) along with silicon and sulphur, tends to be brittle and has a low tensile strength capacity. Cast iron is not readily weldable since it is subject to hardening and cracking in the heat affected zone. It was used extensively for columns in buildings built in the early to middle 19th century and continued to be used in the early part of the 20th century. However, wrought iron became the more dominant material in the late 19th century.

3.2.2 Wrought Iron

Wrought iron became a major structural material following the invention in 1784 by Henry Cort of a process, carried out in a puddling (stirring in the molten state) furnace, involving several stages, including hammering and rolling. Wrought iron is appreciably more ductile than cast iron, consisting of iron interlayed with slag and other impurities. The slag impurities, forming a thin grain within the metal, tend to resemble grain in wood. Carbon content was normally less than 2.1%. This improvement in the quality of iron significantly increased the use and production of iron. However, the through thickness properties are very low making it difficult to weld although welding endwise is feasible since applied stresses are parallel to the grain. Wrought iron starting being used in buildings in 1855, with the manufacture of wrought iron "I" beams. Buildings were limited to five or six stories until the latter part of the 19th century. The use of wrought iron framing with masonry infill was introduced in 1881 and enabled a number of buildings to be built up to 19 stories. Modern wrought iron now has a carbon content of not more than 0.15% with some slag.

3.2.3 Steel

Blacksmiths in medieval times used a process known as **steeling**, involving case hardening the sides of iron to make a cutting edge. This was carried out by heating the iron to sufficient temperature in a charcoal fire followed by quenching. In the 17th and 18th centuries, a method known as the **cementation process** was used to improve quality. The process involved alternate layers of bars of iron and carbonaceous material heated in a furnace for several days. The product was known as **blister steel** due to the blisters on the surface of the bars. Benjamin Huntsman developed, in the 18th

Century, a process of melting blister steel in small fireclay crucibles with a flux material protecting the steel from atmospheric gases. Subsequently the steel was then cast. Around 1856, Henry Bessemer in England and William Kelly of Kentucky, USA developed the process of passing air through cast iron in order to remove the impurities, including carbon, phosphorous and sulphur, from the molten metal to produce steel (see Figure 3.3). The steel was subsequently cast and significant quantities of this material were made in the 18th and early 19th centuries, primarily for small products such as tools, instruments, etc. Further improvements were made by the addition of alloying elements. The first specification for structural steel, establishing quality control, was published in 1894 – 95. Standardization of shapes was agreed by manufacturers and steel proceeded to dominate the structural market. By the early 20th century, steel beams of varying sizes and widths, up to 61 cm (24 inches) in depth, were being manufactured.

3.3 STEEL PRODUCTION

3.3.1 Processes

The development of the **Bessemer process** (see Figure 3.3) around 1856 substantially changed the use of steel. The process involved changing liquid iron to steel by blowing air through it to significantly remove the carbon from it by oxidation creating carbon monoxide. Silica based refractory was used to line the vessel. In 1876, Sidney Gilchrist Thomas improved lining of the vessel, by using chalk stone, in order to make steel from iron obtained from high phosphorus ores.

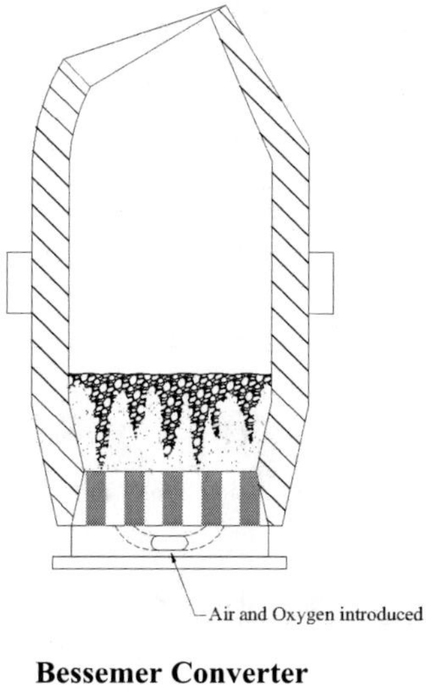

Bessemer Converter

Figure 3.3

The **open hearth process** (Figure 3.4) was first developed by Carl Wilhelm Seimens in 1857. The open hearth process utilizes a shallow steel making area called a "hearth" in which molten iron, limestone and scrap steel are charged. The process involved reducing carbon and other impurities by burning them out of the iron. In order to develop the high temperatures, gas and air are preheated and are drawn through the chambers built of brickwork which absorbs the heat. The flow through the furnace is reversed and the gas and air are heated by the brickwork. This cyclic process enables the furnace to reach temperatures of 1650°C (3000°F) sufficient to melt the iron. Later developments of the process included oxygen lances to assist with reducing impurities. At first, the Bessemer process was favored because it was shorter in duration. However, the open hearth process became more favorable because the steel it produced had more uniform chemical and physical properties.

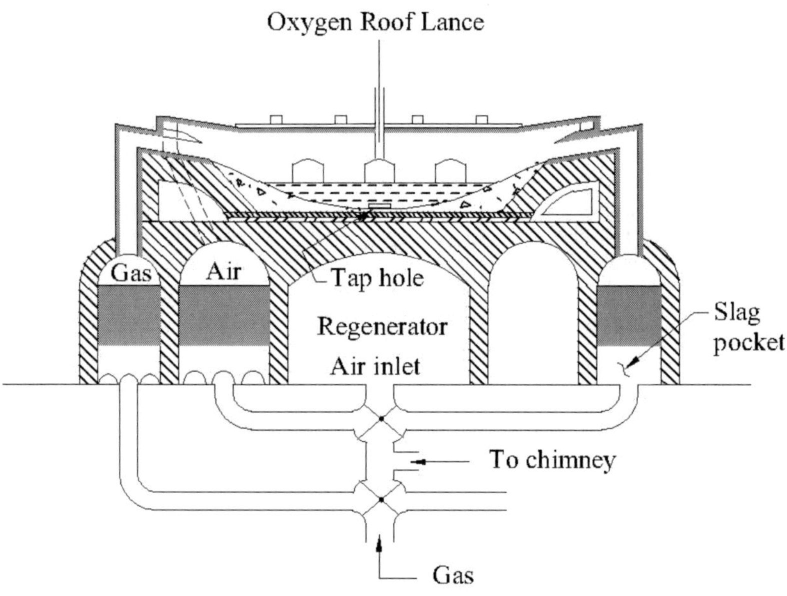

Open Hearth Process

Figure 3.4

The open hearth furnace replaced the Bessemer process in the United Kingdom and in the USA although the Bessemer process was used in Europe for a long time. After 1890, steel prices fell by 50% and steel became the dominant material in the ship building and in the building industries.

The **Electric Arc-Furnace** (EAF), involving heat being generated from electric arcs, (see Figure 3.5) was first introduced around 1880. Heat generated by electric arcs struck between carbon electrodes and the metal bath was used for high alloyed steels, stainless steel and heat resisting steel. Oxygen lancing was used for removing carbon. For several decades it could not compete on production costs with the open hearth and Bessemer processes.

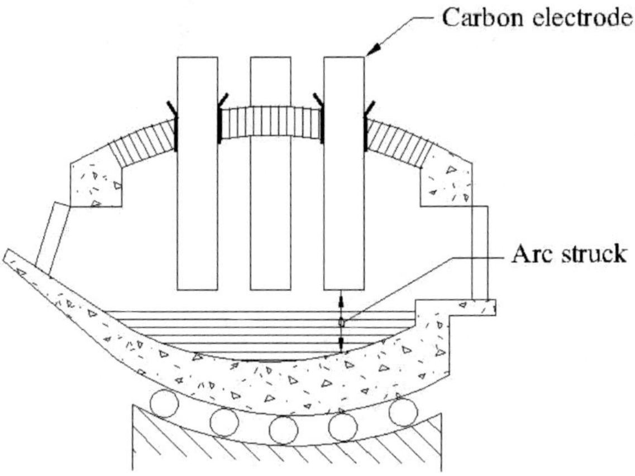

Electric Arc Process

Figure 3.5

In Europe, the **Thomas process**, which adopted the Bessemer process but also used nitrogen, was substantially used until just after World War II. The steel was subject to strain age embrittlement which contributed before World War II to cracking during fabrication (see Chapter 1, Section 1.1.4). Strain age embrittlement occurs when iron and nitrogen form iron nitride which tends to reduce ductility and toughness. Also, the iron nitride tends to be attracted to regions where significant internal stresses occur. The Thomas/Bessemer process can now be considered extinct. After World War II, increased availability of oxygen and further development of blowing oxygen from the top led to the improvement of the **Linde Frankel** process (first developed in 1928) in Austria. The oxygen is blown on to the top surface of the liquid steel which circulates until the carbon content is reduced (see Figure 3.6). The process is also known as the LD process, named after the cities of Linz and Donawitz in Austria where it first went into commercial production in the early 1950's. It is also known as the basic oxygen process (BOP).

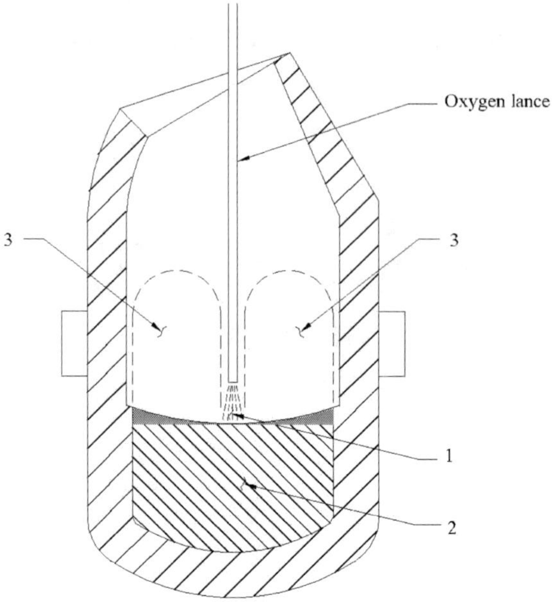

Refining in the LD vessel takes place in three regions:
1 - Oxygen penetrates the slag and reacts with liquid metal producing local temperatures up to 2500 °C (4530°F).
2 - Circulating of the bath from the hot spot.
3 - Slag metal reactions.

Linde Frankel (LD) Process

Figure 3.6

Eventually, the EAF process became cost effective such that presently all structural shapes in the USA and a significant proportion of foreign producers are produced through melting in EAF. The EAF process is more capable of using scrap than other processes. The EAF burden comprises four major components:

- Steel scrap
- Alternate or Supplementary iron units
- Alloying elements
- Slag Formers

The furnace is first charged with the burden. Carbon electrodes are then lowered into the furnace and heat is generated when an arc is struck. The charge is melted relatively quickly. The primary melting energy is the electric arc assisted by chemical energy. Pure oxygen is blown into the furnace to "burn" oxidizable elements (e.g. carbon, manganese, silicon, aluminum). Sacrificial carbon is then added to react with excess oxygen. Steel producers have a high degree of control over the carbon content in steel and a varying degree of control with the elements contained in the steel.

3.3.2. **Ladle Metallurgy and Casting**

In traditional steel making, liquid steel is poured from the ladle into a series of cast iron molds. It is then cooled in the cast iron mold and solidified into an ingot. Solidification rates are affected by the liquid steel temperature, ingot size and steel chemistry. As steel solidifies (or freezes) at a certain temperature, elements and metallic compounds in the steel tend to segregate. Pure iron is usually the first to segregate followed by sulphur, phosphorous, carbon, silicon and manganese. If steels solidify rapidly, segregation has less time to occur. The slower the rate of solidification, the greater is the rate of segregation. Segregation porosity and shrinkage cavities (piping) occur during the solidification process and are dependent upon composition, temperature and ingot size (see later for further discussion).

Since the 1980's, steel productivity has been substantially increased by the development of the **continuous casting process** (see Figure 3.7). The

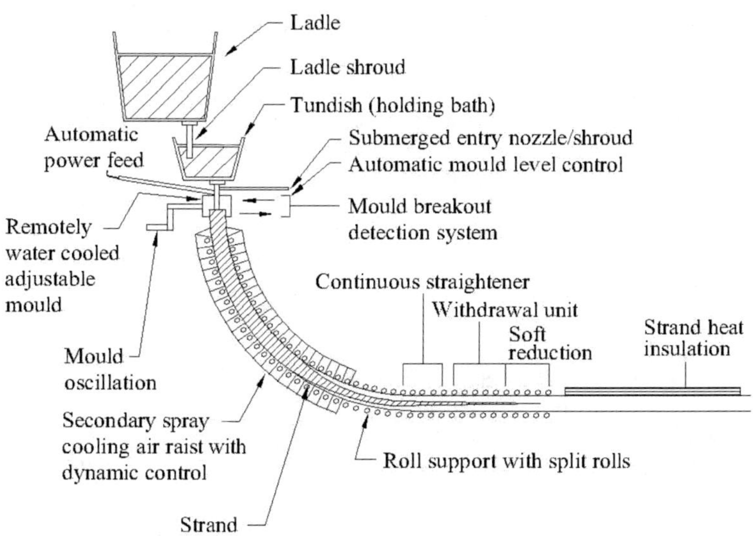

Continuous Slab Casting

Figure 3.7

continuous casting process involves "killed" liquid steel (deoxidized steel, see later for more discussion) being fed from ladle to a holding bath called a **tundish** and then to the adjustable water cooled mold. Liquid steel, in contact with the water cooled mold made of copper, quenches such that a thin layer of steel on the surface solidifies, known as a solid shell, is formed. Within the mold the shell is such that it is capable of maintaining its cross section and containing liquid steel at its core. Partially solidified metal, called a **strand**, first descends vertically, conveyed by water cooled rollers and eventually emerges horizontally. Outside of the mold, shell thickening continues with the help of water and/or air sprays. All steel is clean, with high productivity and little waste. However, the process does have segregation problems where carbon, manganese and phosphorous migrates to center of slab. Secondary treatments such as dephosphorization, vacuum treatment, calcium injection and quenching can help minimize migration.

Methods of improving the quality of steels include the following:

Desulphurization

High sulphur content can have an adverse effect on steel properties. The basic approach to remove sulphur involves sulphur reacting with calcium or magnesium, injected into the liquid metal, to form stable sulphides which do not significantly affect the properties of steel. Desulphurization decreases as the temperature decreases such that long times may be required to drive sulphur levels lower. Further improvements can be made by using a desulphurising slag comprising lime, fluorspan and a deoxidant.

Deoxidizing Steel

Liquid steel contains dissolved oxygen. Continuous cast steel must have steel with oxygen content reduced such that carbon monoxide is not formed during solidification. The process of reducing oxygen is called killing which is accomplished by the addition of highly oxidizable elements such as silicon and/or aluminum. Steels which have not had dissolved oxygen reduced to prevent carbon monoxide occurring during casting are called **semi-killed** or **non-killed steels**.

Alloying Elements

The required steel chemistry is then obtained by the addition of alloying elements such as manganese, vanadium, columbium and niobium. Inert gas is injected or stirred to more rapidly dissolve the alloying elements, also assisting desulphurization and uniformity of steel temperature and composition.

Segregation and other defects

As steel solidifies (or freezes) at a certain temperature, elements and metallic compounds in the steel tend to segregate. Pure iron is usually the first to segregate followed by sulphur, phosphorous, carbon, silicon and manganese. If steel solidifies rapidly, segregation has less time to occur. The slower the rate of solidification, the greater is the rate of segregation.

Shrinkage at the top of the ingot, leading to significant internal defects, sometimes occurs in killed steels. The defects generally do not heal during subsequent rolling.

Segregation, if present in a wide flange member, is in the shape of a dog bone as shown in Figure 3.8.

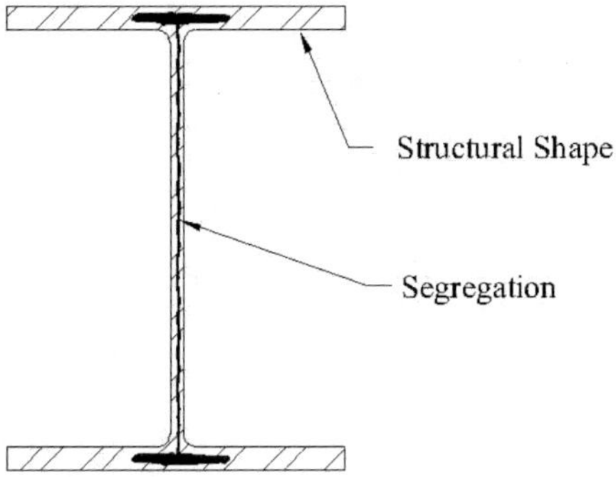

Schematic Representation of Structural Shapes from Ingots
(From FEMA 355A)

Figure 3.8

Inclusion shape control

This involves the addition of rare earths or calcium, zirconium and titanium that form **globular** (spheroidal) **sulphides.** Other metallurgical effects may not be desirable. **Misch metal** is often used to improve inclusion shape control. Inclusion shape control is important to improve ductility, fracture toughness and through thickness properties.

3.3.3 Effects of Thermal History

As described in Section 3.1, the thermal history of steel during the solidification process has a significant influence on the properties of steel. When steel first partially solidifies at a temperature of 1493°C (2720°F), it is known as delta-iron ($\delta - iron$) and upon further cooling to about 1400°F (2550°F), the atomic arrangement changes to gamma-iron$(\gamma - iron)$, the structure of which is known as austenite. Further cooling causes a transformation from austenite, which is a face centered cubic structure, to ferrite, which is a body centered cubic structure $(\alpha - iron)$. Excess carbon is rejected and forms iron carbide known as pearlite. Pearlite strongly increases hardness, strength of steel but reduces ductility.

If steel temperatures are rapidly lowered such that diffusion of carbon does not occur, bainite or martensite (see Section 3.1 for definitions) will form. These products cause steel to be harder, stronger but less ductile.

The process of tempering, which is a controlled raising of the steel's temperature, will cause carbon, trapped in the martensite, to diffuse to produce bainite or pearlite. Ductility and toughness are improved and are accompanied by a reduction in strength and hardness.

Steel when heated to a temperature between 127°C (260°F) and 160°C (320°F), depending on chemical composition, increases in yield and ultimate tensile strengths by as much as 25% above normal temperature values. However, ductility and fracture toughness are reduced.

The yield and tensile strengths of steel, at temperatures above approximately 160°C (320°F), begin to decrease such that, at a temperature of approximately 220°C (430°F), the physical properties are about the same as at normal temperature. At 650°C (1200°F), the steel decreases in volume (shrinks) as it changes its molecular structure $(\alpha - iron\ to\ \gamma - iron)$. The yield and ultimate strengths become very low with no strength at around 1200°C (2200°F).

3.4 ROLLING PRACTICE

With structural steels, the main objective is to produce fine grained steel since reducing grain size lowers the ductile brittle transition temperature, improves toughness and increases yield strength. When the steel is rolled, plastic deformation takes place due to atoms slipping along planes and the material is work hardened. The presence of heat can appreciably modify the tendency for work hardening during rolling. The temperature at which rolling occurs, which can significantly affect steel properties, may be summarized as follows:

- Deformation above 1000℃ (1830°F) produces course grain.
- Deformation carried out between 900℃ (1650°) and 1000℃ (1830°F) can lead to fine grains.
- Deformation carried out between 840°C (1540°F) and 900℃ (1650°F) causes austenite to form elongated grains.
- Deformation carried out below 840℃ (1540°F) can increase the brittle to ductile transition temperature.

Consideration of the above is used in steel mills to control desired steel properties.

3.5 CARBON & FERRITIC ALLOYS

The elements in steel including alloys and impurities may be summarized as follows:

Carbon

Carbon (up to 0.8%) is the main element in the formation of steel causing hardness and strength to be increased with increasing content. The greater the carbon content, the more difficult the steel is to weld.

Increase in Carbon also lowers the transformation temperature and increases Martensite (brittle structure). Higher carbon content tends to increase the risk of hydrogen induced cracking (H.I.C.) due to welding and reduces fracture toughness. Carbon increases strength and decreases ductility and weldability. It both controls the maximum attainable hardness and contributes substantially to hardenability. Carbon has a moderate tendency to segregate.

Manganese

Manganese, when combined with low sulphur content to form manganese sulphides, tends to reduce solidification cracking occurring at elevated temperature. Manganese also strengthens steel by solid solution hardening and grain refinement, which leads to increased fracture toughness.

Chromium

Chromium improves steel quality including higher toughness in the heat affected zone and higher hardness.

Niobium

Niobium raises the recrystalization temperature which helps to give a finer grain structure.

Phosphorous

Phosphorous tends to increase strength and hardness but decreases ductility and toughness. It is considered as an impurity but sometimes is added for atmospheric corrosion resistance. It has a strong tendency to segregate.

Titanium (nitride)

Titanium provides a means of controlling grain size at rolling.

Vanadium

Vanadium works at lower rolling temperatures [around 700°C (1290°F)] to control grain size thus increasing strength and toughness.

Boron

Small amounts of Boron (0.0005%) increases hardenability and is used only in aluminum killed steels (see Chapter 4, Section 4.1.2(ii), Flux Core Arc Welding on hardenability of welds). It is effective with low oxygen and nitrogen steels to lower transition temperature.

Copper

Copper is used in structural steels for weathering resistance.

Molybdemum

Molybdemum improves corrosion resistance and improves resistance to solidification cracking.

Nickel

Nickel improves strength and toughness.

Silicon

Silicon (usually around 0.35% but can be much lower) is primarily a deoxidizing and scavenging agent. Silicon tends to reduce ductility.

Sulphur

Sulphur is primarily an impurity which significantly reduces the fracture toughness of steel. Its content needs to be kept low. One of the reasons for adding manganese is to form manganese sulphides improving the quality of the steel.

3.6 HEAT TREATMENT OF STEEL

Heat treatment tends to relieve internal stresses and remove coarseness of grain and can have significant influence on the properties of welded material. Types of heat treatment are described as follows:

Annealing

The purpose of annealing is to soften the steel, relieve internal stresses and reduce the coarseness of grain. Annealing comprises heating steel to close to the upper critical temperature (phase change from δ-iron to γ-iron, see Section 3.1). Typically the material is heated to above 670°C (1240°F) for many hours (24 – 30 hours) and then slowly cooled (4 – 5 days) usually in a furnace to allow the material to be stress free. Annealing can provide stress relieving but is not usually effective in improving fracture toughness.

Normalizing

Normalizing consists of heating the steel to approximately 38ºC (100°F) above the upper critical temperature. The process is similar to annealing, with regard to heating rates, except the material is heated to 870°C (1600°F) and the material is then cooled in still air. Since the steel material is cooled more rapidly than the annealing process, it results in greater strength but less ductility. According to Stout (1987) heating steel to 870°C (1600°F) and controlling cooling (similar to the annealing process) can significantly improve fracture toughness characteristics and reduce nil-ductility temperatures.

Quenching

The process of quenching is used to provide hard steels. The steel is heated to above the critical temperature and held for sufficient time to change the structure. It is then quenched in oil (sometimes water) such as not to change the microstructure of the steel. The rate of cooling is important to establish desired properties and can be affected by the alloy content. The resulting hardened steel can be very brittle.

Tempering

Tempering is carried out after quenching to reduce the brittleness of steel. The steel is heated to below the critical temperature, held, then cooled slowly to provide the required properties. Tempering is carried out in oil, salt or lead baths and also in furnaces controlled by fans circulating air. The combination of quenching and tempering can produce the best combination of strength and notch ductility. For further discussion on the effects of thermal changes, see Section 3.3.

3.7 TENSION & HARDNESS TESTS

Tension tests are usually carried out on small circular or rectangular specimens and are carried out at a slow rate (see Section 3.12 on discussion on loading rates). The specimens are machined such that the section effectively tested has a smaller area than the end segments. The end segments are threaded to be engaged in grips. In the U.S.A., ASTM Standard A370 is normally applied for tension testing.

Hardness tests, of which there are several types (Scleroscope, Brinell, Vickers, Rockwell) typically involve impact applied to the steel. Although essentially these tests are to determine the ability of the material to deform, attempts at correlating the data with ultimate tensile strength have been made.

The **Scleroscope Hardness Test** involves a hammer and a rounded diamond tip dropped from a fixed height. The hardness is determined from the rebound height and measured according to the Mohs hardness scale. Very hard steels have a Mohs hardness of about 7.

The **Brinell Hardness Test** involves a steel ball, 10mm (3/8 inch) in diameter, applied with a large force. The Brinell Hardness number (HB) is determined from the applied force divided by the curved surface area of the indentation. Typical HB numbers for steels vary from 200 (soft) to 500 (hard).

The **Vickers Hardness Test** is similar to the Brinell test except that the indenter is in the shape of a pyramid with a diamond point. The Vickers Hardness Number (HV) is obtained from the applied force divided by the surface area of the pyramid like depression. Vickers Hardness Numbers for steel are similar to Brinnell Hardness Numbers.

The **Rockwell Hardness Test** utilizes a cone shaped diamond point on steel ball. Different sizes are used for different materials. In this case, the depth of the indentation is measured.

Although hardness tests have been used to estimate ultimate tensile strengths, they can also be used to assess variation in steel properties and potential for steel cracking due to application of heat when welding.

3.8 STEEL PROPERTIES

Steel, tested with either a circular or a rectangular cross section, displays an elasto plastic behavior. In the elastic range, the strain is fully recoverable. As is well known, the rate of increase of stress with increase in strain is called Young's modulus. As the load is increased, the strain becomes nonlinear and permanent plastic deformation occurs (see Figure 2.17a) eventually resulting in necking of the specimen. Normally, increasing stress is required to produce increasing strain causing **strain hardening**. Yield strength is generally measured at 0.2 percent strain. The rate at which stress increases with plastic strain is called the **Strain-Hardening Modulus**. The steel hardens to a peak value (ultimate tensile strength) then usually decreases until the specimen fails. If a specimen is unloaded, after being strained into the strain hardening region and is then immediately reloaded, it returns to the stress strain curve. However, the ductility at fracture will be reduced by the magnitude of residual strain.

If the specimen is left for several days until it is reloaded, it may return to a stress strain curve above the original curve resulting in higher tensile strength. This phenomenon, which is known as **Strain Aging,** also results in increase in strain hardening but with decrease in ductility. Although, usually tensile and compressive strengths of steel are about the same, as mentioned in Chapter 2, Section 2.3.3, when steel is first deformed in tension, then deformed in compression, the compressive strength is lower than if only tested in compression. This property, known as the Bauschinger Effect, also occurs when compression is first applied and when the load is reversed into tension.

As mentioned previously, the varying processes of making steel can have significant influences on the properties of steel. The processes causes grain refinement and elongation of grains in the rolling direction resulting in anisotropic properties (unequal properties in at least two directions) particularly with regard to ductility and fracture toughness. Thus the steel properties, with regard to ductility and fracture toughness, in the transverse or through thickness direction, can be significantly less than for the longitudinal direction.

Some variations in properties with regard to structural shapes, include the following:

- Mill Test Results (MTR) higher than minimum specified ASTM A36 up to 338 MPa (49 ksi) [SSPC (1994)].

- Notable variation in yield strength, fracture toughness and ductility across rolled sections. Beedle and Tall (1959) reported yield strengths 4-7% higher in the web than in the flanges and similarly Galambos and Ravindra (1978). Engelhardt et al (1996) reported significant strength variations within 89mm (3.5 inches) from the center of the web. Byefield and Nethercot (1997) also reported higher yield strengths obtained from the web due to finer grain structure and higher carbon content than the flanges. Bartlett et al (2003), regarding the mechanical properties of

ASTM A992 steel, found that the mean ratio of flange yield strength to web yield strength to be 0.953 and consistent with the findings of Galambos and Ravindra (1978). It is interesting to note that Withey (1928) found similar characteristics of property variation in the 1920s.

- Tests reported by Sarkinnen (1998) showed low ductility (only 11%) and low toughness in through thickness tests. Sarkinen's results are illustrated in Figure 3.9. These results indicate the anisotrophy of steel in terms of ductility and toughness particularly with regard to through thickness properties.

- Significant variation has been found in the "K" area region of some wide flange members. This is discussed in Section 3.9.

- Low toughness in Hollow Steel Sectionsat the seam weld and corner regions. This is discussed in Section 3.10.

- Significant increases in yield strength occur due to higher strain rate. See Section 3.12.

The effects of temperature on yield and tensile strengths are described in Section 3.3. It should be mentioned that the modulus of elasticity also decreases significantly commencing at a temperature of about 100ºC (212ºF).

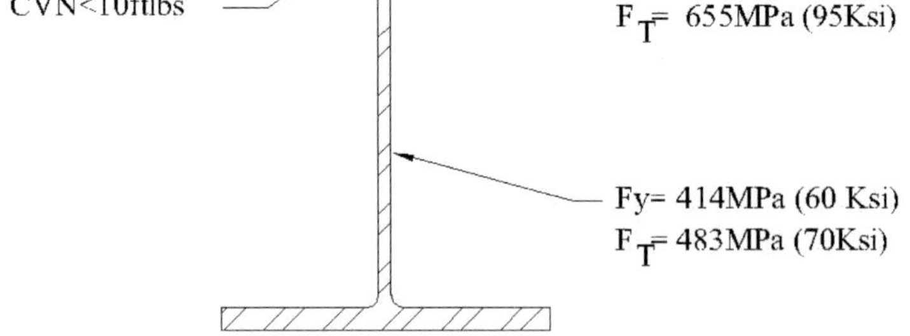

Wide Flange Member Properties

Figure 3.9

3.9 TOUGHNESS & DUCTILITY IN WIDE FLANGE MEMBERS

The K area is defined as the region of the web that extends from the tangent point of the web and the flange-web fillet [K dimension in AISC (2005)] to a distance of 38mm (1½ inches) into the web beyond the K dimension (see Figure 3.10). Around 1995, concern for low toughness and ductility in the K area of wide flange members was expressed by the industry. During fabrication on some projects, involving welding of continuity and/or double plates in wide flange members at moment frame connections, fractures occurred. Some fractures originated at the toe of the fillet, between the section flange and web extending into the web area. Some full scale tests on welded moment resisting beam column connections for the SAC project (see Chapter 1, Section 1.2.11) resulted in failures due to fractures running along the K area. Similar fractures were noted in some wide flange columns of existing buildings following the 1994 Northridge earthquake. [Maranian (1997)].

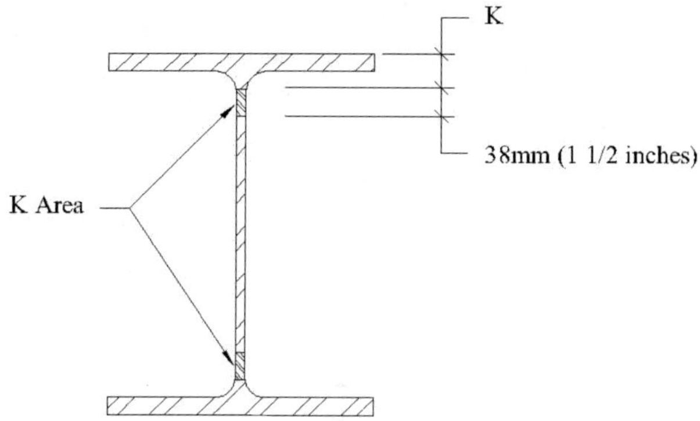

K Area

Figure 3.10

The significant differences in the properties at the K area and other areas of the section apparently are due to cold roller straightening (known as rotary or roller straightening) of the sections. Cold roller straightening is necessary as wide flange members, after cooling down, often have bows that exceed [ASTM A6]. The Rotary straightening method is carried out continuously and the rollers apply cold working which also includes the area between the web and flanges of wide flange members. The contact stresses cause the mechanical properties in the K area to become stronger but less tough and less ductile.

Rotary or roller straightening is used primarily for lighter members typically less than 223 kg/m (150 lb/ft). Heavier sections use the gag straightening procedures which involves deforming each member as a simple beam and not imposing significant contact stresses.

Issues on the K area may be summarized as follows:

- Low ductility
- Low toughness
- Prevalent in members under 223 kg/m (150 pounds/ft.)

Gag straightening, which is typically used on heavier members, gives improved properties. This is because gag straightening is applied to short lengths of the member and does not tend to work harden the K area.

Application of controlled heating and quenching can improve the properties and some steel wide flange members are produced with these supplementary processes.

AISC (2005) requires Charpy Vee Notch tests (CVN) to be carried out in the core area as shown in Figure 3.11 in order to ensure adequate toughness properties in this area. This is required for columns in seismic lateral resisting systems in the United States [AISC (2005)].

It was first recommended by Yee et al (1998) to keep the welding of the stiffener (continuity) plates well away from the K area. This is now an AISC requirement for Seismic Design (AISC Seismic 2005).

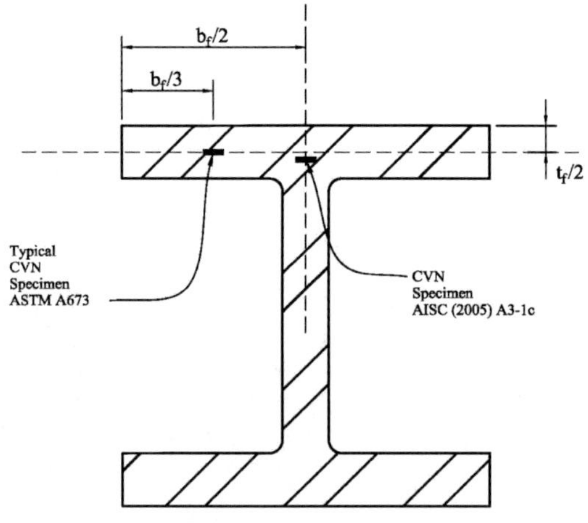

Charpy Vee Notch Specimen
Locations Specified in ASTM A673 and AISC-LRFD

Figure 3.11

3.10 TOUGHNESS IN HOLLOW STEEL SECTIONS

There has been concern for low toughness in hollow steel sections (HSS) particularly following the Northridge California 1994 Earthquake. As mentioned in Chapter 1, Section 1.1.9, Figures 1.7a, 1.7b and 1.7c from the 1994 Northridge Earthquake show rupture of braces appearing to initiate from the corners. Similar types of failures occurred in tests of a two-story special concentric braced frame using rectangular hollow steel sections carried out at the University of California Berkeley by Uriz and Mahin in 2004 [SEAOSC (2005)]

A steel tube post being erected on a project in Alaska fractured along its length when hit with a sledge hammer when being erected (see Figure 3.12). The cause of the cracking is unknown. However, low temperature and low toughness in the corners of the tube appear likely causes. This failure appears similar to the fractured girder that occurred in Belgium in 1934 (see Chapter 1, Section 1.1.4). Figure 3.13 shows a crack in a tube supporting a stair discovered several years after construction. The circumstances of the cracking are not known.

Kosteski et al (2005) carried out Charpy V-Notch (CVN) testing on HSS sections manufactured in North America, South America (Brazil) and Europe (France, Germany and Finland) primarily to assess variations in toughness characteristics in the sections following concerns for an advisory given in AWS D1.1 (2004). The advisory in AWS (2004) discusses that for ASTM A500 (cold formed), hollow sections, products manufactured to this section may not be suitable for those applications such as dynamically loaded welded structures, etc., where low temperature notch toughness properties may be important. Furthermore, they advise that special investigation or heat treatment may be required if this product is applied to tubular T, Y and K connections. A similar statement is made in ASTM A500 [ASTM A500 (2001)].

Kosteski et al carried out 557 CVN tests on coupons taken from various locations around the cross section at temperatures from approximately -75°C (-103°F) to 50°C (122°F) to obtain the complete toughness – temperature transition curve for each location. Locations included the weld seam. The HSS sections, from which the CVN specimens were taken, included cold formed (Canada and Finland), cold formed and stress relieved (Canada and France), hot rolled (Germany), hot rolled then cold-shaped (Brazil).

Fractured Tube in Alaska
(Courtesy of Saif Hussain)

Figure 3.12

Cracked Post at Stair

Figure 3.13

Kosteski et al noted that in North America, the prevailing specification, ASTM A500, has no notch toughness specifications for various grades. However, in Europe, cold formed structural tubing requires a minimum notch toughness of 27J (20 ft.lbs.) at -20°C (-4°F) (S335J2H to EN 10219-1 given in CEN 1997). The Canadian Standard (CSA 2004) has CVN toughness requirements for different yield strengths and temperature applications.

Low toughness was found in the corners of the tubes and at the weld seams of cold formed HSS sections made in Canada whereas the cold formed HSS section made in Finland had good toughness properties all around the section. Stress relieving the cold formed sections made in Canada and France did not significantly improve the low toughness characteristics in the corners and at the weld seams. Kosteski et al were of the opinion that the stress relieving temperature utilized, 496°C (925°F) ±14°C (±25°F), is well below the normalized temperature of 830 °C (1530°F) - 900°C (1650°F) and does not produce any significant metallurgical change.

The hot rolled HSS section made in Germany had very good fracture toughness characteristics all the way around the section. The HSS section produced in Brazil, by first hot rolling pipe and then cold forming into square or rectangular HSS section, also had very good fracture toughness characteristics all the way around the section. The process only caused minor degradation of the notch toughness properties.

Considering the work by Kosteski et al, the observed performance of HSS sections following the 1994 Northridge Earthquake and the testing carried out by Uriz and Mahin (see Chapter 1, Section 1.1.9), the ability of cold rolled HSS sections, to be subjected to yielding and low cycle fatigue, appears questionable. Also, referring to Figures 3.12 and 3.13, the use of cold formed sections in low temperature conditions appears questionable.

In 2007, defective seam welds were found in some HSS members according to DSA (2007). The defects included (i) small cracks propagating from the root of the weld to the outside surface, (ii) lack of fusion at the root up to one half of the thickness, and (iii) full depth cracking and splits. A limited number of tests by independent testing facilities on materials imported from China and also that produced in North America has been carried out according to AISC (2007). DSA (2007) recommended a thorough material review, particularly mill certification along with careful examination of seam welds from both inside and outside.

3.11 LAMELLAR INCLUSIONS

Significant investigations into lamellar tearing were carried out by Farrar and Dolby in England during the early 1970s when there were significant concerns regarding the problem particularly in the offshore industry [Farrar et al (1975), Farrar and Dolby (1972)]. These investigations led to a European Standard on testing of steels for through thickness (see later). Non-metallic inclusions tend to get rolled out during the rolling process. The inclusions may end up either globular or film-like. Film-like inclusions tend to be more detrimental than globular inclusions. The nature of the inclusions is somewhat dependent on the steel making process and whether the steel is **killed** or **semi-killed**. As mentioned in Section 3.3.2, the term **killed** refers to removal of oxygen and is achieved by the addition of silicon and/or aluminum which both have an affinity to oxygen.

Semi-killed steels usually have a silicon content less than 0.1%. These may have high oxygen contents and hence a high population of silicate inclusions. The sulphide inclusions tended to be globular and the sensitivity to lamellar tearing is dependent on the size, shape and distribution of the silicates. **Silicon killed steels** have silicon contents of between 0.15% and 0.3%. These can have high oxygen contents but the distribution can vary throughout the ingot with concentrations of sulphide inclusions towards the top and bottom giving rise to laminations. Silicate inclusions are the dominant inclusion type and sensitivity to lamellar tearing is dependant on these and less so on the globular sulphides. **Fully killed steel**, treated with aluminum for grain refinement tend to have low oxygen content which lead to the sulphide inclusions to be film-like. Lamellar tearing sensitivity is influenced by sulphide film-like inclusion content which in turn is influenced by the sulphur content.

For fully-killed steels, where sulphide inclusions tend to be film-like, Lancaster (1992) and Farrar et al (1975) recommend that the sulphur content should not exceed 70 ppm. Barsom and Korvink (1997) warn that the percentage of sulphide content does not yield any information on the shape of inclusions and their distribution and therefore is only a qualitative measure of steel anisotropy. The film-like inclusions, if present, can occur in the flanges of wide flange members. The application of welds tends to cause internal residual stresses which can cause the film-like defect to propagate, commonly known as lamellar tearing. External stresses, due to loading conditions, may also result in lamellar tearing. The proportion of the crack tends to be a stepped crack pattern as it finds the weakest path (see Figure 3.14). Weld details to minimize lamellar tearing are discussed in Chapter 4, Section 4.1.5.

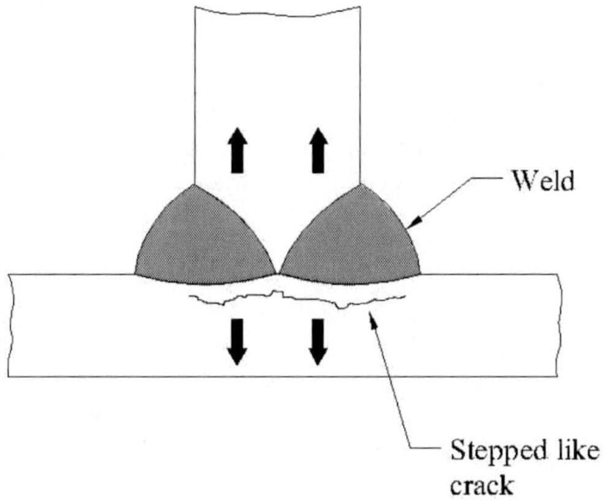

Lamellar Tear / Crack

Figure 3.14

Problems with lamellar tearing still emerge from time to time (see Figures 3.15a and 3.15b). The lamellar tearing occurred in a moment frame connection used in a building in Southern California. The size of the column was a W14x145.

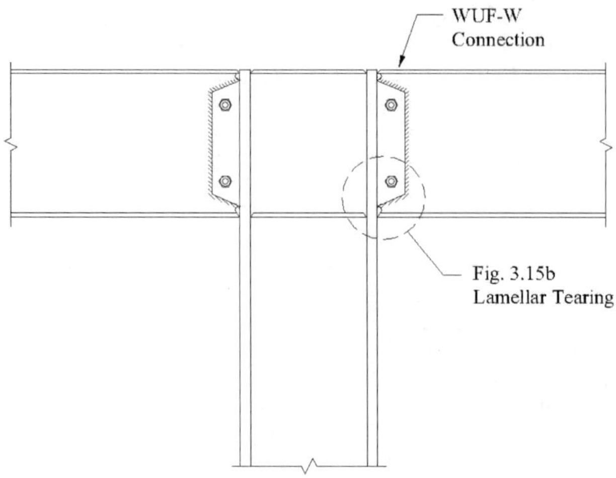

Lamellar Tearing in Moment Frame Connection for a One-Story Structure (Los Angeles, circa 1999)

Figure 3.15a

As mentioned previously, there is a European standard to control steel properties perpendicular to the surface (European Standard 10164). The standard involves carrying out tensile tests on through thickness specimens in the parent metal. These tests check for tensile capacity of the material perpendicular to the grain of the steel.

There are tests available to check for the susceptibility of steels to lamellar tearing [Stout (1987)]. The most common is by ultrasonic testing to determine if filmlike nonmetallic inclusions occur. Other tests include the Cantilever Lamellar Tear Test and the Cranfield Test. The Cantilever Lamellar Tear Test involves alternatively applying weld beads, located in a groove weld between a horizontal plate and vertical plate and inducing load in a hydraulic ram applied to the horizontal plate. Eventually a failure occurs in the vertical plate. The crack is usually parallel to the face of the vertical plate and is step like at its ends. The Cranfield test also involves welding. Successive welds are applied in the acute angle between two plates at 45 degrees to each other. The weld shrinkage of successive passes eventually causes a rotation and a lamellar tear at the root of the test plate. The number of passes to cause a tear provides an indication of the sensitivity to failure.

Photo

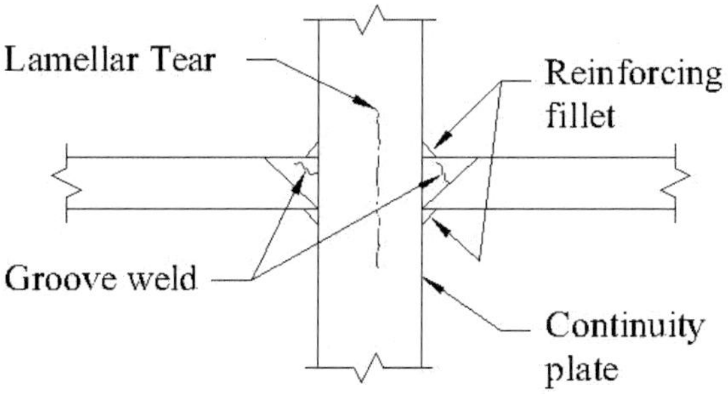

Detail of Joint

Lamellar Tearing in Moment Frame Connection for a One-Story Structure (Los Angeles, circa 1999)

Figure 3.15b

3.12 STRAIN RATE & TEMPERATURE

Tests have shown that steel strengths increase with strain rate. Increases in yield stress can be of the order of 28 MPa to 35 MPa (4 to 5 k.s.i.) at loading rates of 690 MPa (100 k.s.i.) per minute. A loading rate of this magnitude is permitted in ASTM Standard 370 which is used for standard tensile tests. Thus mill test results may be 28 MPa to 41 MPa (4 to 6 k.s.i.) higher than gravity load yield strength. On the other hand, dynamic yield strength, at loading rates of 1 to 2 seconds to maximum load, may be 28 Mpa to 41 Mpa (4 to 6 ksi) higher than mill test values [FEMA 355A (2000)]. Also, yield strength tends to increase with decreasing temperature.

It is also interesting to note that the difference in yield stress between static loading and impact loading (time to fracture ≤ 0.001 second) can be of the order of 172 MPa (25 k.s.i.) [FEMA 355A, (2000)].

Bennett and Sinclair (1965) derived the strain rate – temperature parameter R as follows:

$$R = T \times l_n \left(A / \dot{\varepsilon} \right)$$

Where:

T	=	Temperature (Fahrenheit or centigrade)
A	=	is a material constant (1/second)
l_n	=	logarithm to base n
$\dot{\varepsilon}$	=	strain rate (rate of strain per second)

The R parameter (Fahrenheit or Centigrade) can be used to determine the effect of strain rate on yield stress and tensile strength.

According to Ishikawa et al (1998), if R is the same, even though temperature and strain rates are different, fracture toughness remains constant.

3.13 GALVANIZING STEEL

Galvanizing involves the coating of steel with layers of zinc in order to protect the steel from corrosion. The zinc coating acts as a sacrificial layer that corrodes instead of the steel that it protects. During the galvanizing process, when the steel is immersed in a bath of zinc, a reaction occurs between the iron in the steel and the zinc such that the outer layer is zinc and successive layers are a combination of iron and zinc with pure steel on the interior. To prepare the steel for galvanizing, it is chemically cleaned usually by a process called “pickling” usually using sulphuric acid.

In the pickling stages, a chemical reaction occurs between the chemicals, the rust and the steel which produces hydrogen that is absorbed into the steel. Most of the hydrogen is expelled when the cleaned steeled is immersed into the molten zinc. However, some hydrogen may remain which can cause hydrogen embrittlement. High strength steels, which may be considered to be 790 MPa to 1000 MPa (115 to 145 ksi), can be particularly susceptible and may be subject to brittle behavior.

Also, areas of the steel which have undergonc excessive cold working prior to galvanizing can be affected by the galvanizing process. This is because hydrogen atoms are particularly attracted to areas of high stress and their presence causes additional stresses which can lead to fracture. Furthermore, the heat from the galvanizing process can accelerate strain age embrittlement of the steel.

These concerns can be avoided by the use of mechanical cleaning instead of by pickling. Alternatively stress relieving can alleviate some areas of the steel which have been cold worked.

RECOMMENDED READING – CHAPTER 3

Byfield, M.P. and Nethercot, D.A., (1997); "Material and Geometric Properties of Structural Steel for Use in Design." *The Structural Engineer,* Institution of Structural Engineers, 4 November 1997.

Dowling, Norman E., (1999); Mechanical Behavior of Materials – Engineering Methods for Deformation, Fracture and Fatigue, 2nd ed. Prentice Hall.

Farrar, et al, (1975); Investigations into Lamellar Tearing: A Compendium Reports From a Sponsored Research Programme Phase I-IV and Phase V-VII; The Welding Institute, England.

FEMA 274, October 1997; NEHRP Commentary on the Guidelines for the Seismic Rehabilitation of Buildings.

FEMA 355A (2000); State of the Art Report on Base Metals and Fracture.

Kosteski, N., Packer, J.A. and Puthli, R.S.; Notch, Toughness of Internationally Produced Hollow Structural Sections; Journal of Structural Engineering, ASCE, February 2005.

Lancaster, J; Handbook of Structural Welding, Woodhead Publishing Ltd., 1997 (UK).

Withey, M.O., (1928); Tests of Specimens Cut From Different Portions of Structural Steel Shapes," *Proceedings of ASTM.,* Vol. 28, Part II.

CHAPTER 4

CONNECTIONS AND FABRICATIONS

4.1 WELDING

4.1.1 History

Pressure or hammer welding of gold circular boxes was practiced from the earliest times. Forge welding was used to make artifacts in Syria, Egypt and other areas of the Eastern Mediterranean during the 11th to 9th centuries BC.

Fusion welding was known to be used 2000 years ago during the Bronze age when molten bronze was poured between two bronze parts. This method was typically used for repair of swords and for making chariots by the Chinese. Welding by hammering, which is the art of blacksmithing, was highly developed during the middle ages to make items of iron.

Welding techniques first were used on an industrial scale towards the end of the 19th century following the discovery in England by Sir Humphrey Davies of acetylene and the ability to create an arc between two carbon electrodes. Gas welding, arc welding and resistance welding were all developed during the period 1877 to 1903. Gas welding used oxygen, hydrogen and acetylene and suitable torches and gas storage techniques were invented. Arc welding, using a carbon arc, was developed by Nikalai N. Bernardos and Stanislaus Olsezwaski, who were Russian researchers working in a French Laboratory. They took out a British patent in 1885 and an American patent in 1887 for welding using an electrode holder. Despite the early work by Bernardos and others, arc welding took much longer to develop. At first, bare wires were used which resulted in brittle metal, high in nitrogen. A.P. Strohmeyer, around 1900 in Great Britain, introduced a metal electrode coated with clay or lime. Further developments were made by Oscar Kjellberg of Sweden using stick electrodes dipped in thick mixtures of carbonates, silicates, etc. Resistance boxes, using direct current (DC) power, were used to deposit the electrodes. Wire wrapped with asbestos or paper improved the properties of the weld metal. Further improvements in the wire coating were made using a mixture of metals, ferro alloys and sometimes organic materials bonded with sodium or potassium silicate. C.J. Halslag developed the use of alternating current (AC) for welding around 1919. Both heavy coated rods and light or washed coated rods were tried during the 1920s and specifications for covered electrodes were established in the 1930s.

Resistance welding, originally developed by Elihu Thompson with his patents taken out in 1885, was well established by the 1920's. The process uses spot and seam welding for thin plates and butt welding for such items as chains and bars.

Oxyacetylene welding became a fully developed process by 1916. It was used for welding thin steel, aluminum, copper plates and for repair and maintenance work. The process developed then does not differ very much from the process used today.

Prior to World War I, fusion welding was used primarily as a means of repairing worn or damaged metal parts with connections usually made with rivets. However, during World War I, some of the first all welded ships were built.

Use of riveted connections continued to be the dominant method of joining plates until World War II when fusion welding was used significantly particularly in ship building. It is very difficult to rivet steel more than 51 mm (2 inches thick) whereas welding was found to be capable of achieving connections between thick elements. Welded joints are typically much simpler than riveted or bolted connections in configuration, invariably saving weight.

In the 1920s much research took place in shielding the arc by externally applied gases as it was realized that oxygen and nitrogen in the atmosphere caused brittleness in the steel. Hydrogen was used by Alexandre and Langmuir whilst Hobart and Devers used argon and helium.

Gas tungsten arc welding, which began with work by Coffin in 1890, was further developed in the 1920s by Hobart and Devers and was perfected by Meredith in the 1940s. This led to the establishment of the Gas Tungsten Arc Welding Process. Further work in 1948 included the use of continuously fed wire replacing the tungsten electrode which established the Gas Metal Arc Welding Process. The use of carbon dioxide was developed by Lyubavskii and Novoshilov in 1953 which led to economies in welding.

During the 1950s, a process called "Dualshield" using external gas and flux in the core of the wire was developed. In 1959, the self shielding "Innershield" process using interior and exterior flux was developed by the Lincoln Electric Company and became established as an economic process widely used today.

The electroslag welding process originated with research in the United States by R.K. Hopkins in 1940 and further developed and perfected in Kiev,

Ukraine and Bratislava in the Czech Republic during the 1950s. It was first used in the United States for the fabrication of welded diesel engine blocks.

The improvements in welding techniques in the 20th century has led to the ability to construct complex structures such as space rockets, deep diving submersibles, containment vessels for nuclear reactors, box girder bridges, tall buildings, etc.

4.1.2 **Welding Processes**

Welding is defined by the American Welding Society (AWS) as follows:

> *"Welding is a joining process that produces coalescence of materials by heating them to the welding temperature with or without the application of pressure or by the application of pressure alone and with or without the use of filler metal."*

There are numerous welding and joining processes that have been developed and are available. These processes are grouped by AWS under the following general categories:

- Arc Welding
- Resistance Welding
- Soldering
- Solid State Welding
- Oxyfuel Gas Welding
- Brazing
- Other Welding and Joining

Only a few of the processes developed are commonly used in Steel Structures and are briefly discussed in this publication. Several of the figures shown are similar to figures typically found in books on welding. Discussion on the processes is as follows:

<u>Shielded Metal Arc Welding (SMAW)</u>

This process involves establishing an electric arc between a coated "stick" electrode and the base metal melting the electrode and base metal (see Figure 4.1). The electrode has a coating called a flux which, when heated by the arc, creates a shielding gas that protects the molten metal from the atmosphere (oxygen, hydrogen and nitrogen). The weld metal tends to become covered by slag from the electrode coating. Control of current and travel speed are important to control the size and quality of the weld. Although slower than other processes, the SMAW process is versatile both in the shop and in the field and is capable of producing very good quality welds.

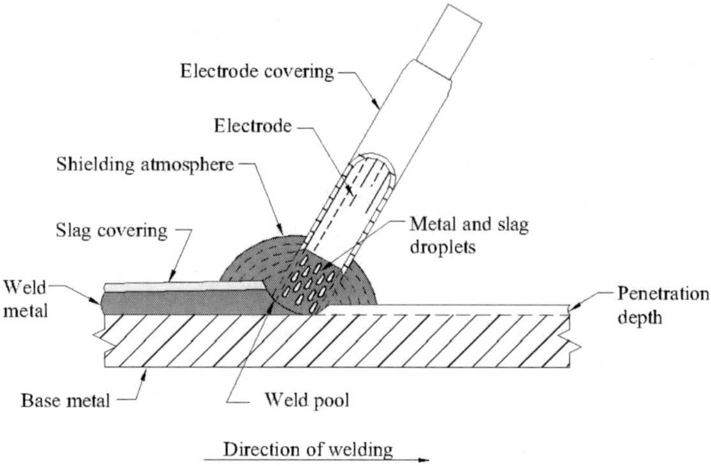

Section

Shielded Metal Arc Welding (SMAW)

Figure 4.1

Gas Metal Arc Welding (GMAW)

This process used externally supplied gas to provide shielding to the continuous electrode and the weld pool (see Figure 4.2). The process is also known as Metal Inert Gas Welding. As with the SMAW process, the electric arc between the electrode and the base metal melts both the electrode and base metal. The shielding gas protects the arc area from contamination and from the atmosphere. The process is versatile, productive, eliminates slag, reduces smoke and fumes and can be used by operatives with a lower skill level when semi-automation is provided. It is usually adopted in the shop and not in the field since the effectiveness of gas shielding can be reduced by wind.

Flux Core Arc Welding (FCAW)

This process uses a flux contained in an interior core within the continuous electrode that provides shielding to the electrode and weld pool (see Figure 4.3). The surface of the base metal and the end of the electrode are melted by the heat between the continuously fed electrode and the weld pool. In some cases, additional shielding is provided by externally applied gas.

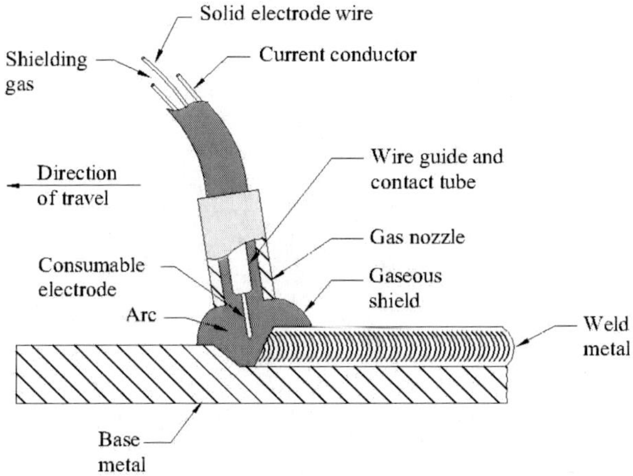

Section

Gas Metal Arc Welding (GMAW)

Figure 4.2

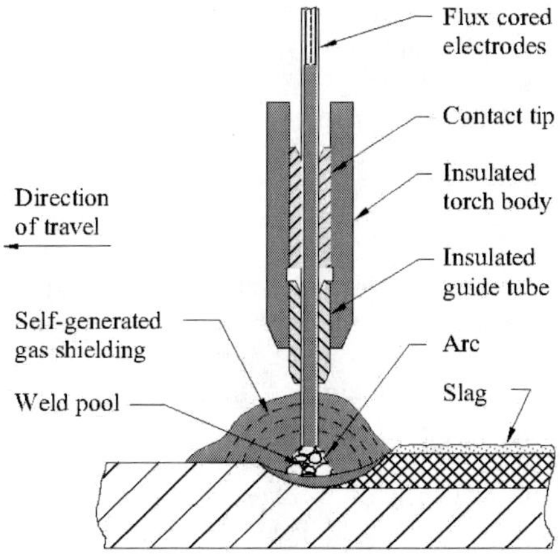

Section

Flux Core Arc Welding (FCAW) with Self-shielded Electrode

Figure 4.3

The type which only uses a flux contained in an interior core within the electrode is called Flux Core Arc Welding, self shielded (FCAW-S). This process has, since the mid to late 1960s, become the process of choice in structural steel applications primarily due to its high productivity.

Flux-cored wires commonly used in the building industry prior to 1994, complied with AWS classification E70T-4 which had no toughness requirement. Following the 1994 Northridge Earthquake, weld metal, using the E70T-4 electrode, was found to have low toughness, apparently due to the use of aluminum, which acts as a deoxidizer. FCAW-S electrodes with significantly improved toughness characteristics, achieved with the addition of nickel, are available.

It is interesting to note that cracks have been encountered in welds made using the FCAW-S process when carrying out welding on a "debarking drum" at a paper mill according to Perdomo et al (2006). In this case, the electrode employed had the AWS classification E70T-1 which requires a Charpy Vee Notch requirement of 27 Joules (20 ft.lb.) at -18° C(0°F). The double vee groove welds were carried out on ASTM A36 material 32 mm to 35 mm (1-1/4" to 1-3/8") thick. Perdomo et al found that the weld samples had a relatively high hardness which they believe was caused by the presence of approximately 0.008% Boron. Boron is known to increase the hardenability in low-carbon steels. Perdomo et al conducted tests on flux-cored wires from five manufacturers. Cracks did not occur when the Boron content in the electrode was less than 0.006%. Perdomo et al recommended that the Boron content should not exceed 0.003%.

Submerged Arc Welding (SAW)

This process uses an arc between bare metal electrode(s), immersed in a blanket of granular flux, and the weld pool (see Figure 4.4). The process, which is automatic, uses the heat of the arc between the continuously fed electrode and the work to melt the electrode and the base metal. The granular flux is first laid directly over the weld area and forms a glasslike slag during melting that floats to the surface. The process is widely used particularly for long seam welding of structural shapes.

The SAW process can cause large heat affected zones. This is due to the high heat input that usually occurs with this process. For further discussion related to the World Trade Center building collapse in 2001, see Section 4.1.4.

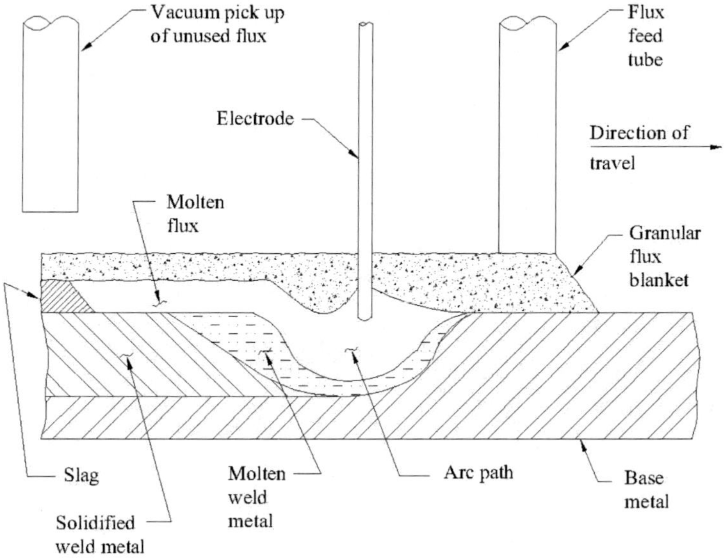

Section

Submerged Arc Welding (SAW)

Figure 4.4

Gas Tungsten Arc Weld (GTAW)

This process uses an arc between a nonconsumable tungsten electrode and the weld pool (see Figure 4.5). A shielding gas, flowing from the nozzle, is used to displace air. Filler metal is not normally used for thin metals and edge joints. Filler metal, externally fed and melted by the arc, is used for thicker sections.

Little or no spatter and smoke occurs. Very high quality and clean welds can be made by this process in any position.

Stud Welding

This process uses a stud gun holding a metal stud (or similar element) which creates an arc between the stud and the base metal followed by forcing the two materials (similar to forging).

When the arc is first established, the heat of the arc melts the stud and the base metal. The stud is then immediately forced against the base metal by

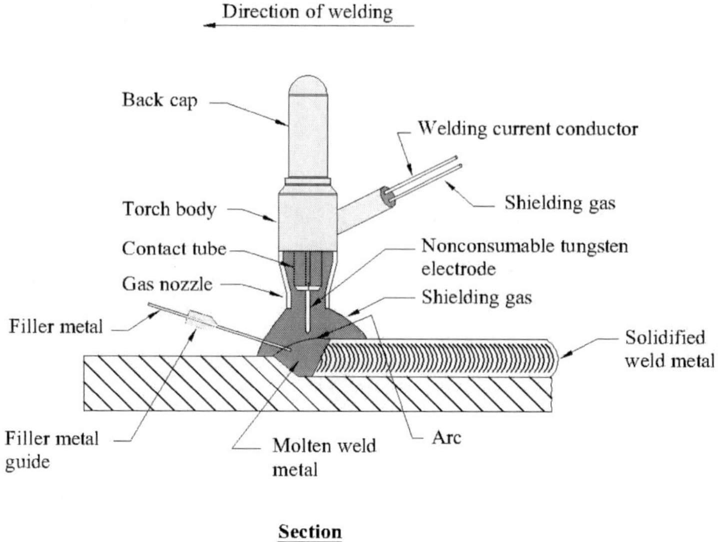

Gas Tungsten Arc Welding (GTAW)

Figure 4.5

the stud welding gun. Ceramic ferrules and sometimes fluxing ingredients placed on the arcing end of the stud provide partial shielding. The ferrules also prevent the ejection of weld metal and assists with reducing glare. Studs are often welded through galvanized sheet steel. The zinc (used in the galvanizing process) becomes volatile in the arc and can cause porosity and fusion defects. This can be offset by increasing the arc time thus removing the zinc before the arc is made.

Electroslag Welding

This process involves an electrode fed automatically through a guide tube commencing at the bottom of the joint where granular flux is also placed creating a layer of slag (see Figure 4.6). The guide tube transmits the current to the electrode. The molten flux (slag), which floats above the molten metal, is electrically conductive and passes the current from the electrode to the base metal melting the electrode and edges of the base metal. The molten metal is shielded by the floating molten flux. Molding shoes are used to confine the molten metal and slag and are usually water cooled to control temperatures. High deposition rates are obtainable along with slower cooling rates. Although large grain growth occurs in the weld

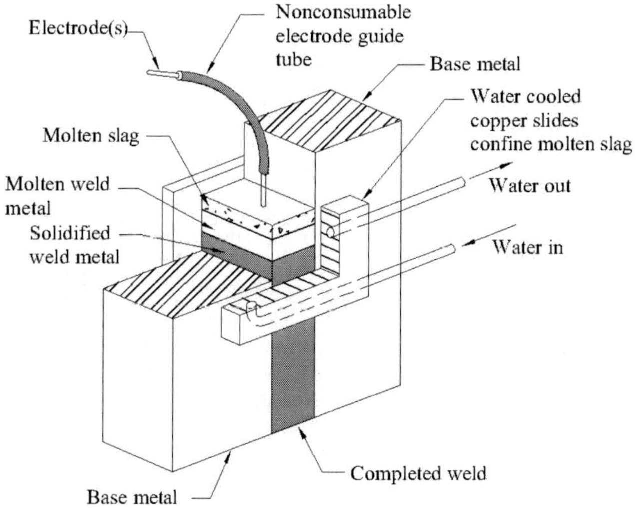

Electroslag Welding (ESW)

Figure 4.6

metal and heat affected zone due to the slow cooling rate, the slow cooling rate can reduce hardness. Before the early 1970's, the electroslag process did not consistently make welds without defects. Adequate toughness was also an issue. This could be improved by normalizing as a post weld heat treatment. More recent developments, including reducing the gap between the surfaces, has appreciably improved weld characteristics.

4.1.3 Joining by Welding

There are three important characteristics of welding using the SMAW, FCAW, GMAW, GTAW and SAW processes that can significantly affect the properties of the welded joint as follows:

Heat Source Intensity

A minimum heat source intensity is needed in order that fusion can be made. However, the heat source intensity needs to be controlled since excessive heat can cause the metal to vaporize. Also, increasing the heat input rate causes the weld pool to be difficult to control and the grain sizes to be undesirably large. Thus a number of passes are required for thick welds using the SMAW, FCAW, GMAW and GTAW processes to control the heat input.

Heat Input Rate

The heat input rate is represented as follows:

$$Heat\ Input\ Rate = \frac{Volts\ x\ Amp\ x\ 60}{Travel\ Speed(in\,/\,minute)}\ joules\,/\,minute$$

The heat input rate controls the heating rates, the cooling rates and the weld pool size. The higher the heat input rate, the lower the cooling rate and the larger the weld pool. Higher heat input rates also allows the grain structure, in the solidifying weld metal, to grow leading to a course structure in the heat affected zone (see Section 4.1.4) and in the weld metal. A course grained structure in steel tends to have reduced ductility and fracture toughness. Therefore, heat input rates need to be limited to control grain size and cooling rates. On the other hand, heat input too low may be inadequate for some processes to make effective welds.

Controlling heat input rate is very important with regard to the quality of the weld and the resulting material properties.

Shielding

At the temperature where electrodes and weld pool have melted, the metal reacts with oxygen and nitrogen in the atmosphere which causes embrittlement. With regards to oxygen, it can combine with iron to form compounds that stay in the weld as inclusions. Free oxygen combines with carbon to form carbon monoxide which can get trapped when the metal solidifies. The inclusions and trapped gases are defects that can cause fracture. In the case of nitrogen, the nitrogen is not very soluble at room temperature. It tends to collect in pockets or forms iron nitrites during cooling and the solidification process. The resulting porosity and inclusions tend to reduce ductility and increase the possibility of fracture. To avoid oxygen and nitrogen embrittlement, the electrode is required to be shielded by flux, gas or both.

The components in electrode flux coatings for SMAW "stick" may come from a number of ingredients including cellulose, metal carbonates, rutile and magnesium silicate derived minerals, calcium flourides, organic material, ferro-alloys and iron powder bonded with sodium or potassium silicate. Each ingredient has an attribute that assists in the welding process. Cellulose and/or calcium fluoride provides the shielding gas. During welding, the drops that fall to the weld pool have slag which coats the weld pool. The slag provides a protective coating. Electrode coating is vulnerable to damage and cannot be used when the electrode is bare. Also, electrodes even left in unopened containers outdoors, will absorb moisture dependent on temperature and humidity. Moisture on the electrodes will introduce hydrogen and oxygen into the weld pool which can increase the potential for cracking. Electrodes are thus required to be stored in heated ovens to remove the moisture.

Shielding, utilizing gas, is carried out through a nozzle. Gases used are carbon dioxide, argon and sometimes oxygen and helium. These gases are also mixed. The composition of the mixed gases can significantly affect welding quality including weld bead contour and weld penetration.

Welding Procedure Specifications

Welding procedure specifications are used to specify the process, joint design details, limitations, procedures and variables to carry out the weld. These requirements are specified in specifications [e.g. AWS D1.1 (2006)]. Variables include voltage, amperage and travel speed to control the heat input (see Section 4.1.3) along with shielding gas type and flow rate (if applicable), preheat and post heat (if required).

Preheat, applied prior to welding by torch or heating elements, reduces the cooling rate, reduces residual and shrinkage stresses, assists in allowing dissolved hydrogen to permeate out without cracking, tends to control fracture toughness properties and reduces hardening at the fusion line.

The level of appropriate preheat is subject to several factors including welding consumables, process, base metal chemistry, thickness of members and environment. Typically, higher preheat is needed for thicker members, where there is high joint restraint, higher carbon content in the welding consumables and base metal, and where the electrodes are not low hydrogen. When applying the same level of preheat, thick plates tend to cool relatively rapidly as the heat is drawn away. However, thin plates cool slowly as the heat does not dissipate. Sophisticated methods are available to determine appropriate preheat levels such as that given in AWS D1.1 (2006), Annex XI. Application of preheating involves the use of gas torches or electrical resistance heaters. When using gas torches, temperature measuring crayons are used to control the preheat temperature.

The **interpass temperature** is the temperature, both maximum and minimum, of the previously deposited weld metal and the adjacent base metal, prior to the start of the next bead, in a multiple-pass weld. Excessive preheat and interpass temperature may decrease fracture toughness and could lead to solidification cracking (see Section 4.1.5). The American Welding Society recommends, for Seismic Resisting Structures, a maximum of 288°C (550°F).

It should be noted that the reheating of each weld pass on the previous pass that occurs during a multi-pass weld tends to refine the grain structure of the previous pass. The resulting network of regions having a fine grain structure tends to enhance the fracture toughness of the weld. Correspondingly, reducing the number of passes, by increasing the thickness of the weld, tends to reduce the fracture toughness of the weld.

Post heat treatment, applied after welding, can reduce the effects of weld shrinkage, strain age embrittlement and hydrogen embrittlement. The post heat treatment, if required, is carried out immediately from the interpass temperature to the post heat temperature. It is then held at this temperature for a time usually dependent upon thickness [e.g. 1 hour per 25 mm (1 inch) of thickness].

Cooling rates can significantly affect the microstructure of the steel. Some areas of the weld and heat affected zone can receive a slow air cooling resulting in a soft structure while other areas can be cooled more rapidly by the cold base metal resulting in a hard structure and less fracture tough. Cooling rates depend on the rate of heat input, the base metal temperature before welding and the section thickness and geometry. Higher heat inputs and preheating tend to slow the cooling whereas heavy sections, which act as heat sinks, cause faster cooling rates. Rapid cooling can result in martensitic structures that have high hardness but can be more brittle. Control of cooling rates can be assisted by the provision of thermal blankets. Electrical resistant heaters, used for preheating, can also be used to control cooling rates.

Excellent discussions on the subject of controlling weld temperature are given by Funderburk (2000) and Stout (1987).

As mentioned previously, preheat is applied partly to reduce residual stresses. Relief of residual stresses can also be achieved to some extent by applying mechanical vibration. The vibrations induced by the mechanical vibrator, with sufficient power and close to resonance, can cause slippage in the microstructure reducing local high yield stresses and thus hardness of the steel.

Peening and hammering can also tend to reduce surface yield point stresses and are sometimes used to improve fatigue life.

4.1.4 Effects of Fusion Welding

Welding Effects

The heat flow during welding tends to affect the composition and micro structure of the weld metal and the heat affected zone. At low levels of current and heat input the filler metal tends to be deposited in drips. With increasing current and heat input, the filler metal flows steadily and the weld pool flow appears to be controlled by electro-magnetic forces. **Undercutting** involves lack of weld and notching at the edge of the weld. This occurs when high current and/or high travel speeds create sufficient electromagnetic force to overcome surface tension and gravity forces such as to displace the weld metal at the sides of the groove. The source of undercutting is often as a result of the technique used by the welder. Left unrepaired, the undercut reduces the thickness and will act as a stress riser possibly inducing fracture. High currents and/or high travel speeds can also cause an undesirable weld profile (sometimes referred to as **humping**) which comprises a series of swellings in the weld. Humping tends to occur in high speed welding using the GMAW process.

There is a potential for absorption of gases into the weld pool. Shielding gases (carbon dioxide, argon and helium) are essentially inert and may only be slightly absorbed into the weld pool. As mentioned previously, their purpose is to shield the weld pool from absorbing atmospheric gases. Nevertheless, some absorption of oxygen, nitrogen and hydrogen from the atmosphere can still occur and may be affected by the nature of the weld pool. Absorption and entrapment of gases can cause **porosity** which can vary from large to small holes. Factors that tend to cause porosity include high sulphur in the base metal, dirty base metal, moisture, oil and the presence of paint. Porosity usually occurs in two forms, i) globular; and ii) blow hole or worm hole or piping. Porosity may occur at the surface and can be visually observed or is buried which can be found by nondestructive testing.

Mechanical Effects (Residual Stresses)

The weld metal is first subjected to compressive stress at high temperature and then weld metal contracts as soon as it solidifies. Left unrestrained, the weld metal contracts proportional to the product of the temperature difference between steel freezing and ambient temperature and the average coefficient of thermal expansion over the temperature range. Significant distortion of members and parts can occur if weld shrinkage is not restrained. Restraint to the weld contraction from the parts to be joined induces significant residual stresses. Residual stresses in thin plates occur primarily in two directions in the plane of the plates. With thicker plates, residual stresses act in all three directions. The stresses can exceed the elastic limit. Since welds may be deposited in several beads, the build up of residual stresses is complex. The first bead of a vee groove weld will be subjected to significant tensile stresses. Sometimes the first bead is susceptible to cracking. The contraction of adjacent passes causes the first bead to go into compression. Further addition of passes can cause the top (last passes) to be in tension, the center in compression and the bottom (first pass) to be in tension again. Figures 4.7 and 4.8 indicate how residual stresses can build up in members.

Figure 4.9, a 30.5 cm (1 ft.) long x 51 mm (2 inch) thick plate tilted nearly 76 mm (3 inches) due to welding, shows distortion due to unrestrained shrinkage of simulated weld repair at the intersection of a horizontal to vertical plate, from Brandow and Maranian (2001).

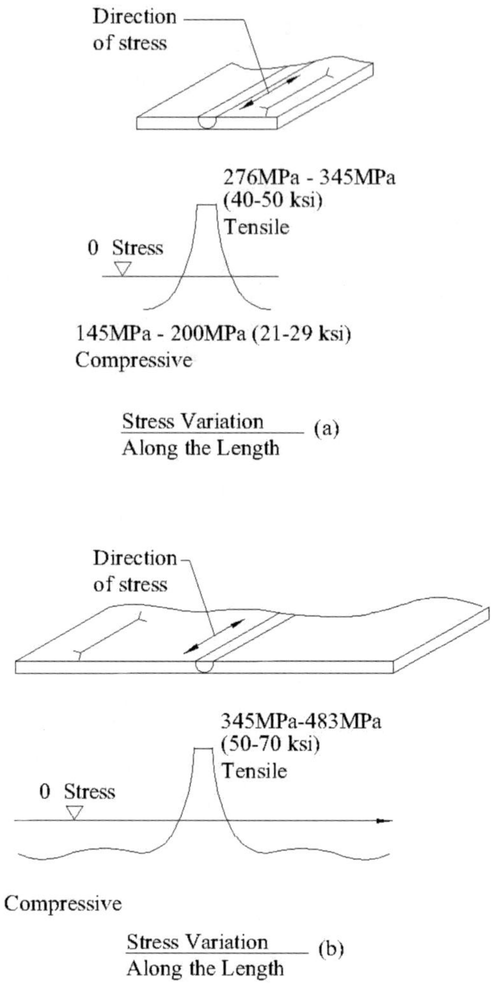

Typical residual stress fields in flat carbon steel plate welded with coated electrodes:
a) Narrow plate (100-200 mm); b) Wide plate (over 300 mm)

Residual Stress at Butt Welded Plates

[Derived from "Analysis of Welded Structures" by Koichi Masabuchi, Pergamom Press, 1980 with permission of Dr. K.Masabuchi]

Figure 4.7

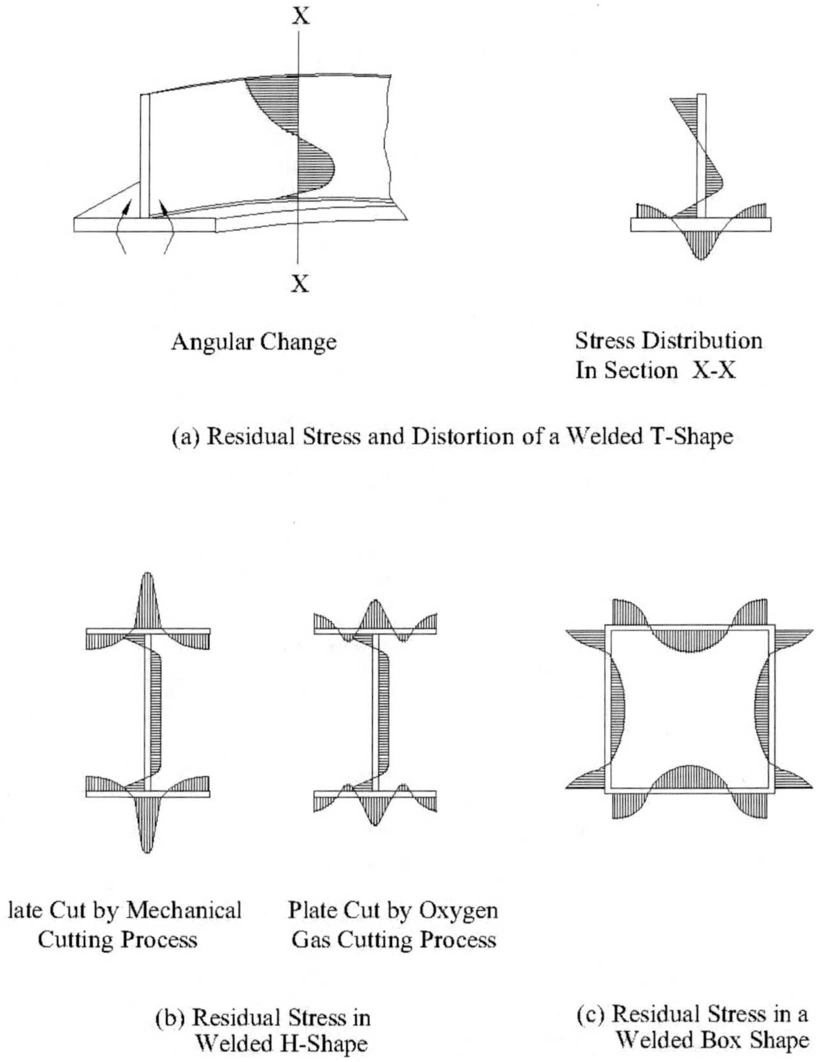

Residual Stress in Built Up Members
[Reproduced from "Analysis of Welded Structures" by Koichi Masabuchi, Pergamon Press, 1980 with permission of Dr. K. Masabuchi]

Figure 4.8

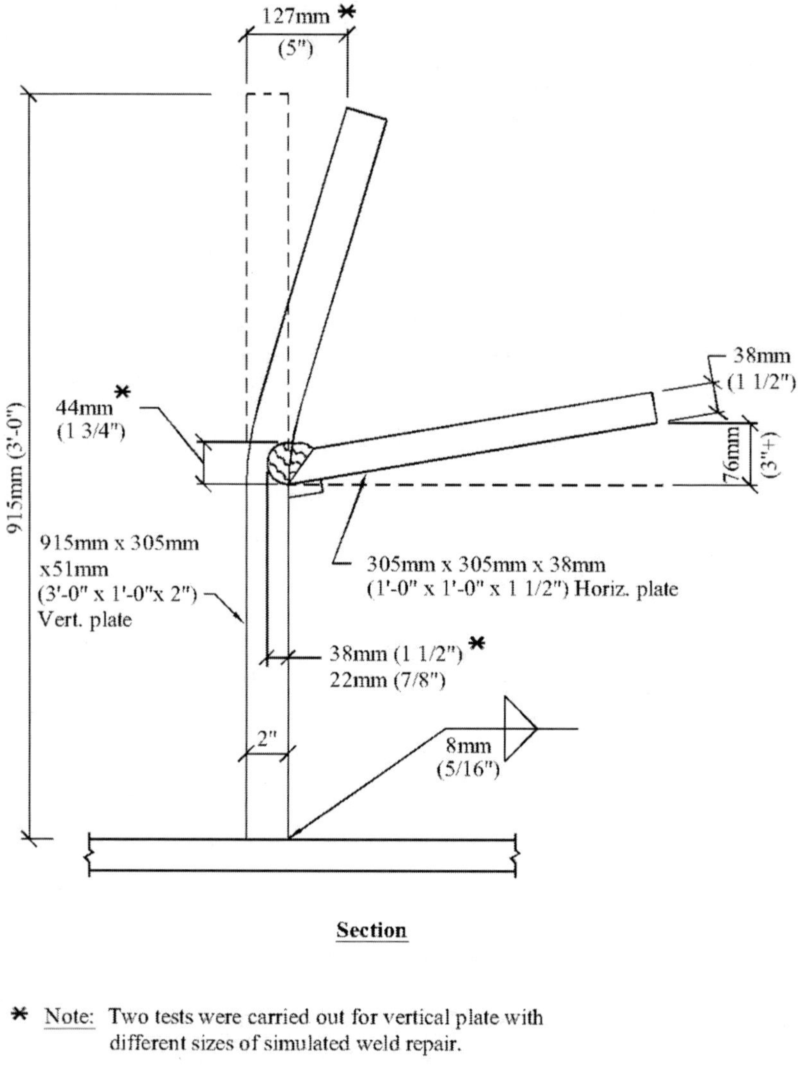

Distortion of Weldment Due to Unrestrained Shrinkage
[From Brandow and Maranian (2001)]

Figure 4.9

Researchers have found that the resulting fracture toughness of weld metal, with residual stresses induced, is reduced. Fig. 4.10, from Dong & Zhang (1998), shows residual stress characteristics which can require the need for higher fracture toughness in order to minimize potential for crack initiation and propagation. Measures to reduce residual stresses include i) use of double bevel welds in lieu of single bevel welds which reduces the amount of weld and enables weld deposition to be balanced; ii) applying post heat; and iii) peening which involves mechanical working of the weld and expands it. Every intermediate pass needs to be peened to reduce distortion. Peening of the first pass should not occur as it may conceal cracks. Also, peening, since it hardens the metal, may reduce toughness and ductility.

Fisher et al (1998) advise that residual stresses due to welding can have considerable influence on the propagation of fatigue cracks in high cycle applications.

A good understanding of residual stresses has been established by several researchers. Residual stresses are much discussed in Campus (1954), Masubuchi (1980) and as previously mentioned Dong and Zhang (1998). Mathematical formulae and computer programs have been developed to assess the magnitude and distribution of residual stresses.

An excellent reference on welding of heavy wide flange shapes, addressing issues including weld shrinkage, is Tsai et al (2001). Tsai et al recommend use of the FCAW welding process, welding the flanges prior to web, increasing the weld access hole to 38 mm (1½ inches) deep. Furthermore, Tsai et al found that weld access holes with a right angle have high residual stresses and recommended tapering the access hole. Similar findings on weld access holes were noted by Ricles et al (2000) which led to the modified access hole shown in FEMA 350 (2000) and AISC Seismic (2005). Ricles et al noted that the best condition was to have no access hole. For additional discussion on access holes pertaining to research associated with surface conditions and varying temperatures, see Section 4.3.5.

Tsai et al (2006) investigated the phenomenon of shrinkage strains, during the weld cooling cycles, causing buckling and distortion primarily on thin shapes. They have established a numerical procedure, based upon the heat input and plate thickness, to determine the peak temperature below which buckling does not occur.

Effects on Heat Affected Zone (HAZ)

The heat affected zone (HAZ) is the area of the base metal immediately adjacent to the weld that is affected by the thermal cycles occurring during welding (Figure 4.11). The microstructure across the HAZ varies. The area immediately adjacent to the weld tends to have a course grain and may have a

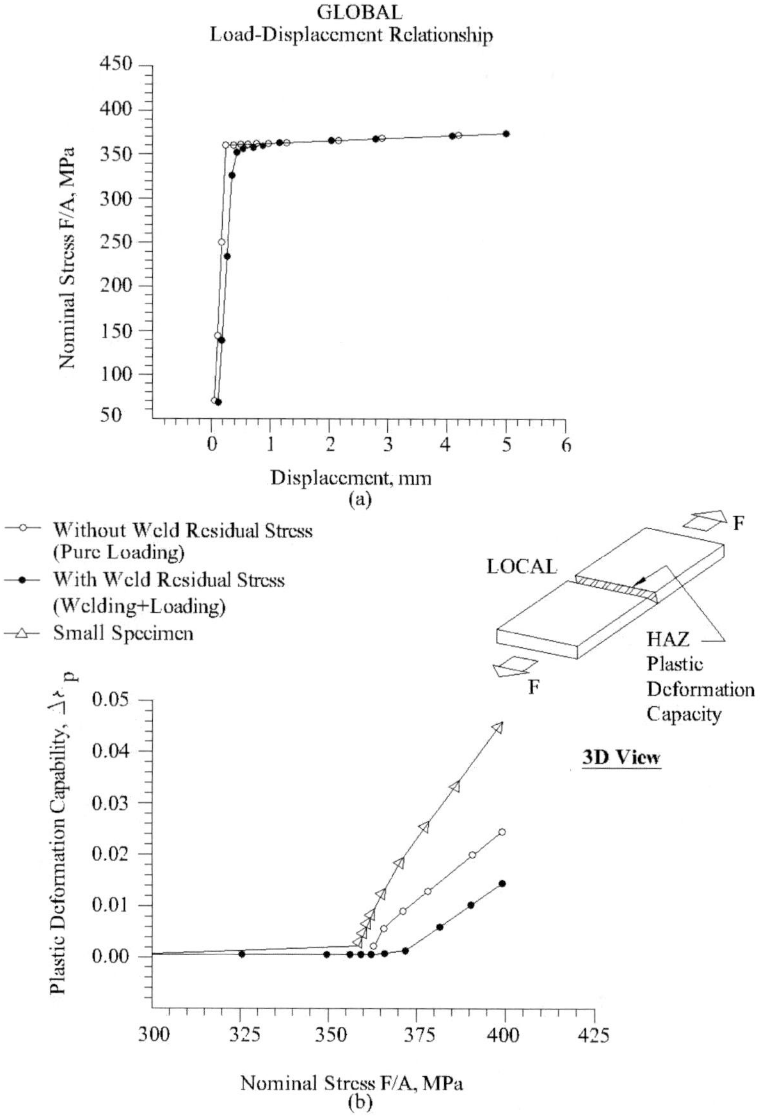

Residual Stress Effects on Structural Behavior

[Dong & Zhang (1998) with permission of the American Welding Society, Miami Florida]

Figure 4.10

martensite structure, which is strong but brittle, and usually occurs due to rapid cooling. The remaining area has smaller grain growth and usually produces a grain size less than that of the parent metal. Figure 4.11 [derived from Patel (2006)] shows varying regions in the heat affected zone. The narrow zone immediately adjacent to the solidified weld metal fusion line consists of an unmixed and partially melted zone followed by a coarse grain structure HAZ (CGHAZ), then a fine-grain HAZ (FGHAZ), an intercritical HAZ (ICHAZ), tempered HAZ, and finally unaffected base metal.

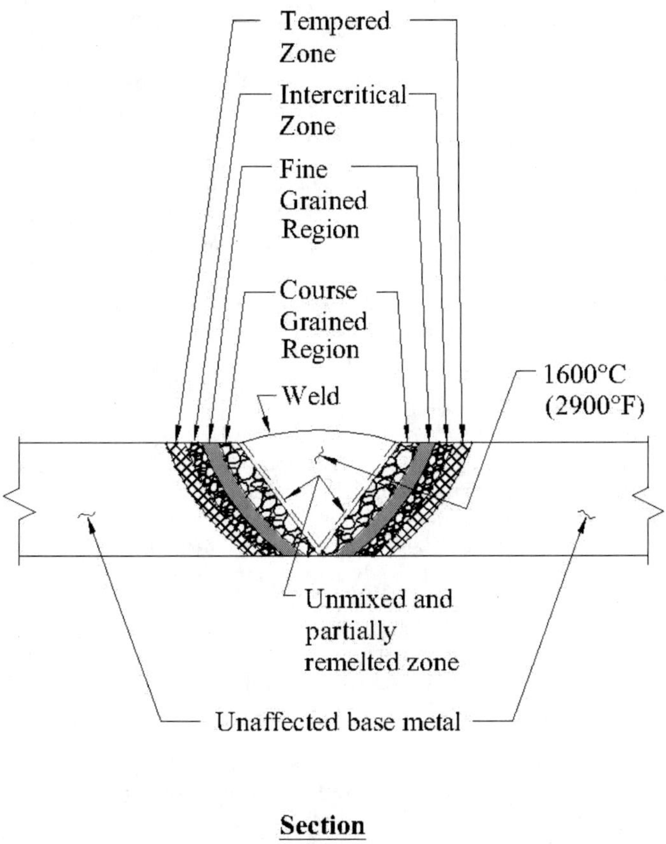

Heat Effected Zone (HAZ)

Figure 4.11

Embrittlement of the HAZ can occur if the steel is heated to above 1200°C (2200°F) then cooled rapidly. Above 1400°C (2550°F) severe loss of toughness can occur and reheating during subsequent passes can further cause embrittlement. Cracking can occur in the heat affected zone particularly with longitudinal loading, where the strain levels are similar to the base metal. The HAZ is also very sensitive to Hydrogen Induced Cracking (see section 4.1.5) and is affected by residual stresses. The most susceptible microstructure is close to the fusion boundary.

Measures to improve the properties of the HAZ are primarily to reduce the heat input and reduce the cooling rate. The less heat applied, the smaller is the HAZ. Heat input can be controlled by the Welding Procedure Specification as described earlier. The slower the cooling rate the less tendency there is for martensite to occur. Cooling rates can be controlled using preheat. That is, heating the base metal prior to welding. Application of sufficient preheat, by controlling the weld metal/HAZ cooling rate minimizes or avoids transformation to hard brittle martensite. Post heating and insulation, subsequent to welding, will also reduce cooling rates. Heat affected zones can be checked for hardness. HAZ with Brinell hardness (see Chapter 3, Section 3.7) of 250 rarely have problems with cracking. HAZ with Brinell hardness in excess of 350 can be vulnerable to cracking. Welds and/or HAZ, not exhibiting cracking, may do so under the application of load particularly cyclic loading. Tack welds, normally involving small fillet welds, without preheat applied, may be subject to high cooling rates resulting in hardening of the weld and HAZ. Ductility is significantly reduced which may result in cracking when stresses are applied.

Methods available to assess the sensitivity for cracking in the HAZ include the Hydrogen Control Method and the Hardness Control Method. The Hydrogen Control Method only addresses the possibility of cold cracking due to hydrogen embrittlement. The Hardness Control Method offers a means to preclude an excessively hard HAZ. Based upon the thickness of the material, the carbon equivalent (derived from the chemistry of the steel) and the maximum allowed hardness level, the minimum critical cooling rates can be determined. Cooling rates should not exceed these critical rates if cracking is to be prevented. Excellent discussions on preheat and HAZ are given in Patel (2006) and Patel (2007).

An interesting investigation, which included much discussion on the heat affected zone area, was carried out by Banovic and Siewert (2006). Their investigation pertained to the failures associated from the aircraft impact damage on the exterior columns at the 97th, 98th and 99th floors of the World Trade Center, Tower 1 which collapsed on September 11, 2001 in New York. Banovic and Siewert were able to carry out metallurgical investigations on samples from the actual columns that were impacted by the airplane (Figures 4.12a and 4.12b).

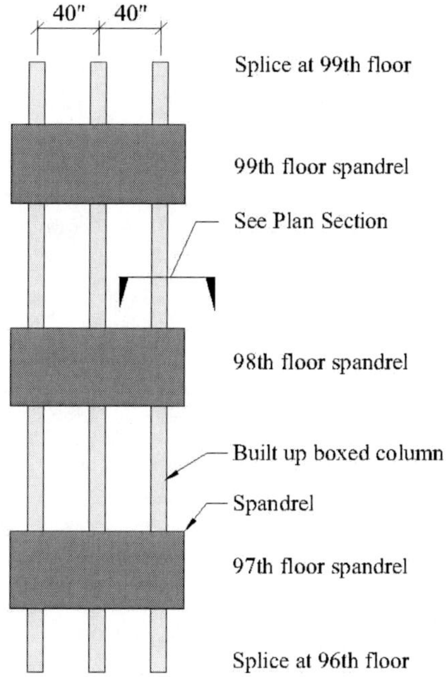

Elevation of Typical Exterior Panel

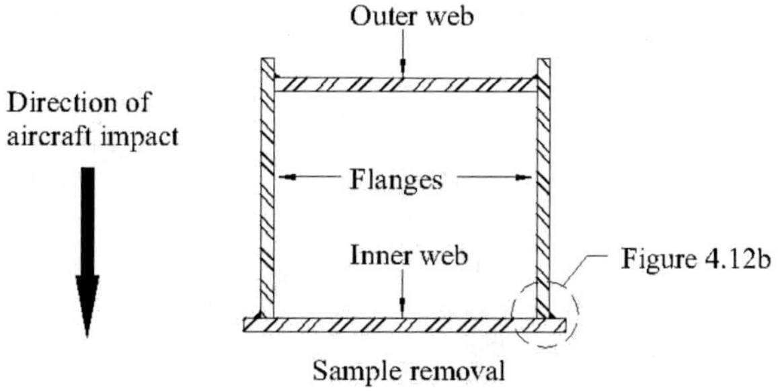

Plan Section

World Trade Center Tower One - Exterior Panel
[Banovic and Siewert (2006) with permission of the American Welding Society, Miami, Florida]

Figure 4.12a

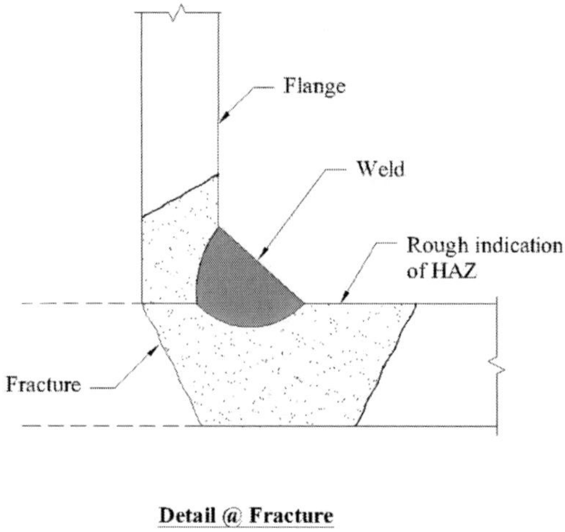

World Trade Center Tower One - Fracture in Panel
[Banovic and Siewert (2006) with permission of the American Welding Society, Miami, Florida]

Figure 4.12b

The columns, typically at 1m (40 inches) on center, were fabricated as part of a three story, three bay panel which were then connected by bolting and welding to adjacent panels during erection of the building. The columns were box shaped built up sections [approximately 35.6 cm (14 inch square)] comprising outer and inner webs (which were parallel to the face of the building) connected to the flanges with typically19 mm (¾ inch) fillet welds. These fillet welds, which were carried out by the submerged arc process (SAW), were exposed on the outside. Spandrel plates at the floors were butt welded to the inner web plates which occurred between the floor lines. The yield strength of the hot rolled steel, which was specified to various strengths depending on the location in the building, was 380 MPa (55 k.s.i.), at the levels where the impact occurred. The steel appeared to have quite good properties and the non metallic inclusions, typically occurring in hot rolled steel, were found to be well dispersed. The impact, which occurred at a very high strain rate, caused fractures which included the intersection of the inner web and one of the flanges in the heat affected zone occurring in the inner plate behind fillet weld. The fractures displayed only localized ductility such that the absorption of the energy of the aircraft's impact energy by the steel components was small. According to Banovic and Siewert, the inner plate had a lower cross section compared with the weld throat. Also, there were no visible flaws or lack of fusion in the welds. The heat affected zones in the inner plate and flange plate, which extended over a significant area, had diminished properties with regard to ductility and toughness. This was indicated by substantially higher hardness compared with the base metal particularly at the interface of the heat affected zone and the weld. Banovic and Siewert were of the opinion that appropriate welding procedures and joining materials were adopted for the general service conditions of the building and that the fractures that occurred were due to the extreme event.

4.1.5 Cracks in Welds

Solidification Cracking

Solidification cracking, sometimes known as **hot cracking**, occurs at temperatures near the melting point during or immediately after welding.

Solidification cracking occurs as a result of contraction of the weld metal inducing tensile residual stresses sufficient to cause fracture of the weld metal due to lack of ductility. The cracks that occur are intergranular fractures (see Chapter 2, Section 2.2.1). Conditions where cracking can occur are shown in Figures 4.13 and 4.14. The crack is usually characterized by a blue appearance along the crack due to surface oxidation occurring at a high temperature. Factors that can affect the weld metal having inadequate ductility include heat input, cooling rate, the material composition, the presence of nonmetallic inclusions and the plastic strains that have already occurred as a result of significant residual stresses.

Applying preheat to control the cooling rate and use of a low heat input will result in a finer grain structure improving the fracture toughness properties of the steel thus reducing the potential for fracture. Special alloys, in the filler metal, such as nickel, again tend to improve fracture toughness. The presence of nonmetallic inclusions acts as defects causing stress concentrations. Use of low hydrogen electrodes can significantly help to minimize the potential for cracking in the weld.

Hydrogen Induced Cracking (H.I.C.)

Hydrogen is derived generally from moisture during welding and is then absorbed into the molten weld deposit. Upon solidification, hydrogen tends to be rejected over time. Atomic hydrogen diffuses out. However, hydrogen, which has changed to the molecular form, migrates to nonmetallic inclusions and areas of high strain. This process is sometimes known as **hydrogen embrittlement**. It then creates high internal stresses which may cause cracking sometimes known as **cold cracking** or known **as delayed cracking**. These internal stresses, caused by the molecular hydrogen, reduce the amount of strain capacity in the material before failure occurs.

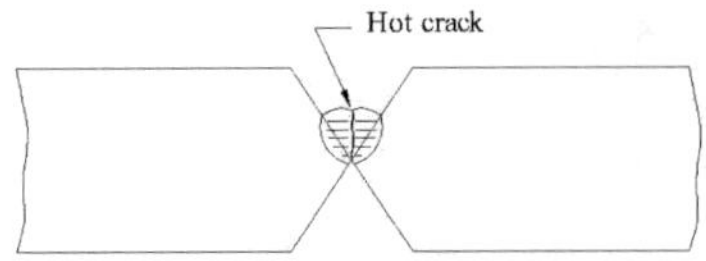

Figure A Hot crack in butt weld root pass

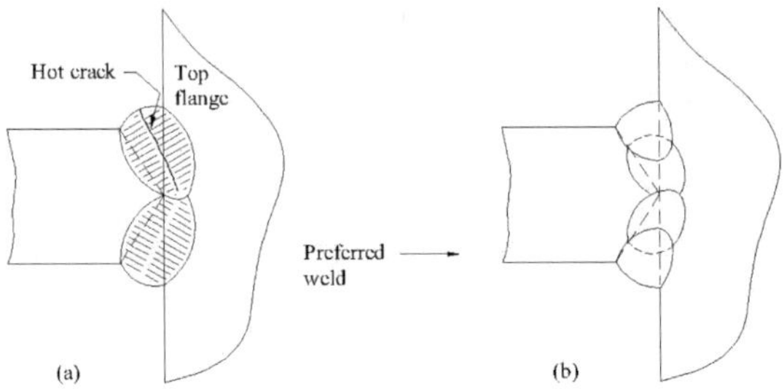

Figure B Full penetration weld of top flange to web

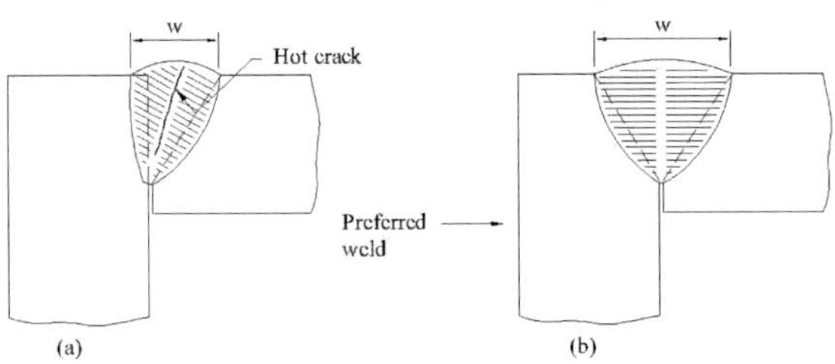

Figure C Box girder corner weld

Solidification (Hot) Cracking

Figure 4.13

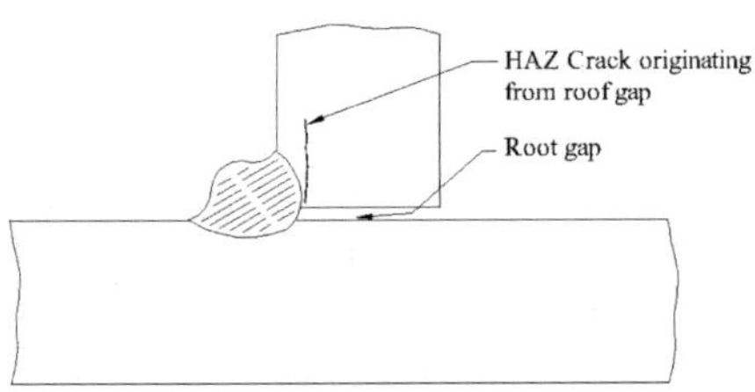

Root gap HAZ crack

Solidification (Hot) Cracking

Figure 4.14

The shielding atmosphere which, although mostly contains carbon dioxide, also includes hydrogen gas derived from organic elements and moisture that are used to bind the ingredients of the flux coating or core.

Primary considerations for the potential for cracking due to hydrogen embrittlement include the hydrogen content, the fracture toughness of the weld, heat affected zone and parent metal and the stresses imposed by welding and subsequent to welding. Cracking tends to occur usually within days although sometimes it can occur after several months. According to Pargeter (2003), delay times in cracking tend to increase with increase in heat input and increase in yield strength. Hydrogen embrittlement is reversible upon the removal of the gas and material ductility can bc partially restored. The risk of cracking tends to increase with increase in section thickness due to the greater distance for hydrogen to diffuse from the mid sections.

Measures to reduce the potential for hydrogen induced cracking include material with good fracture toughness properties, use of low hydrogen electrodes, design to avoid restraint, control of welding procedures and use of pre-heat and post heat treatment. Although less beneficial for control of grain growth, welding procedures with high heat inputs are less susceptible to cracking than welding procedures with low heat inputs. Furthermore, maintaining preheat temperature after welding has commenced has been found to give significant benefit. This can be done by maintaining a minimum interpass temperature of 149°C (300°F) [Lazor et al (2005)]. Recommendations on delay times for inspection for hydrogen cracks are given by Pargeter (2003).

Lamellar Tearing

As mentioned in Chapter 3, Section 3.11, lamellar inclusions, filmlike non-metallic inclusions, present in the steel, can propagate due to welding causing internal residual stresses or from externally applied loads (see Figure 4.15). The fracture, which is brittle, usually occurs close to the heat affected zone and typically steps as the fracture progresses from one inclusion to the next. The work by Farrar and Dolby (1972) led to the development of recommended weld details to minimize the potential for lamellar tearing. These were also published in AISC's Journal 1973 [AISC (1973)] and are incorporated in AISC's Steel Construction manual [AISC (2005)] (see Figure 4.16a).

A testing program, investigating the short transverse fatigue properties of structural steel, was carried out at the University of California, Berkeley by C.J. Adams during the mid-1970's [Adams (1975)]. Adams carried out cyclic testing on "tee" specimens comprising of two transverse plates welded to a longitudinal plate with complete penetration double bevel welds. The plates were 38 mm (1-1/2 inch) thick with a yield strength of 392 MPa (56.9 ksi) and

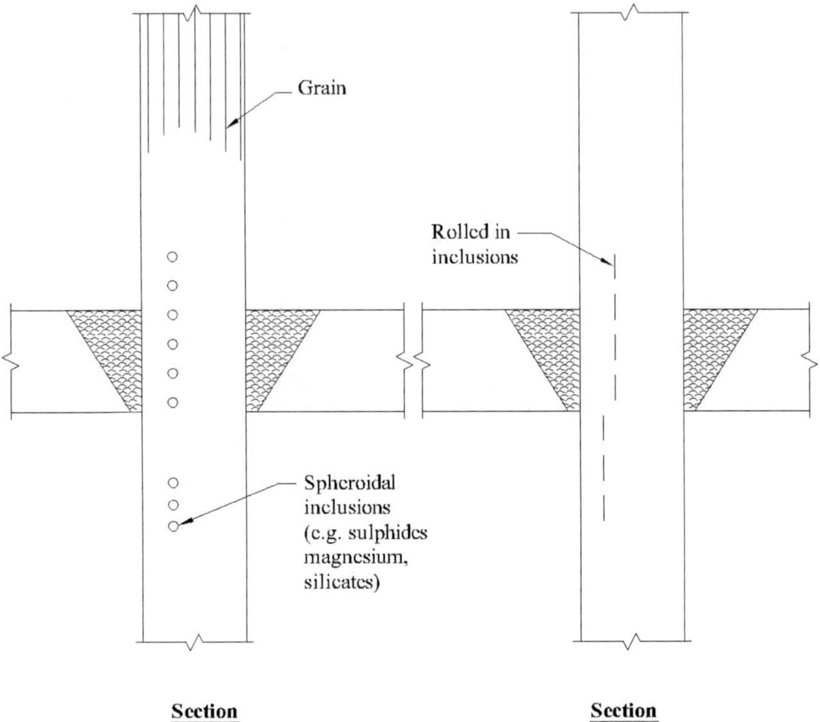

Lamellar Inclusions

Figure 4.15

a tensile strength of 532 MPa (77.2 ksi). Preheat was used for welding, and although no lamellar tears were detected by ultrasonic testing, some hot cracks, adjacent to the heat affected zone (HAZ) were subsequently found to be undetected. The 76.2 cm (30 inch) long assemblies were cut by a band saw such as to make 63 specimens assemblies. The specimens were machined into four short transverse types representing adjacent to the HAZ, through the HAZ, and through the weld on each side of the longitudinal plate. Specimens representing the longitudinal direction were also made. These specimens were cyclically tested for axial loading with varying axial loads to determine the S-N curve (see Section 2.3.2). Adams found poor low cycle fatigue in the short transverse direction. In comparison, the longitudinal direction had as much as 12 times the life of the short transverse specimens under low cycle fatigue. For high cycle fatigue, Adams found even greater superiority of the

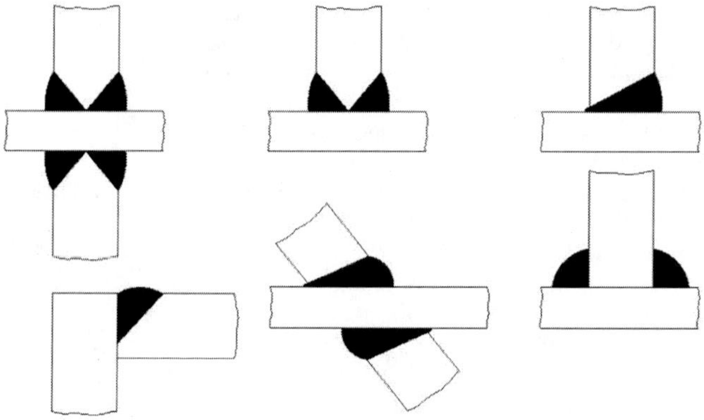

Typical large weld details that may lead to lamellar Tearing

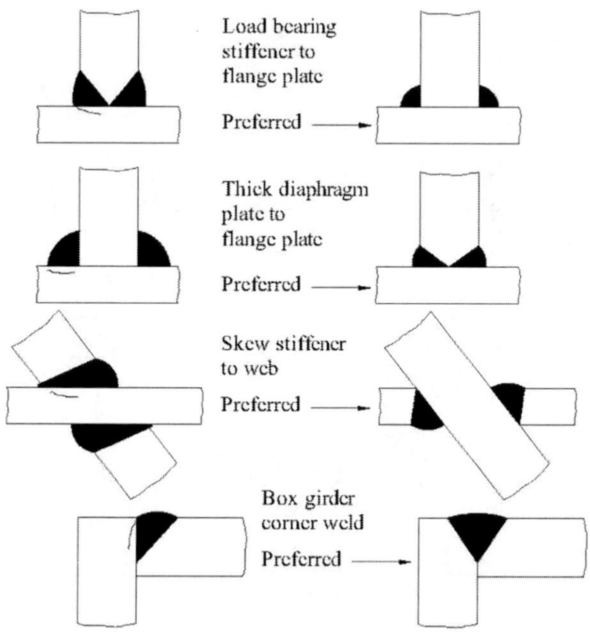

Preferred weld details

Lamellar Tearing

Figure 4.16a

longitudinal direction over the transverse direction than for low cycle fatigue. Adams found the transverse direction (through thickness) to be very unreliable and lamellar tearing to be a primary concern. The importance of detecting lamellar tearing and repairing when found was emphasized by Adams. Furthermore, Adams found the weld material significantly more resistant to fatigue than the base metal. Adams recommended the weld yield strength to be less than the yield strength of the center of thick plates.

Although the SAC project (see Chapter 1, Section 1.2.11) had concluded that lamellar tearing was not an issue, other researchers Lindley et al (2001) were of the opinion that, for heavy sections with sulphur contents in the range of 0.016 to 0.020%, lamellar tearing in the column flange adjacent to the weld could occur during seismic loading conditions. Lindley et al stated that, for applications where very high restraint is likely to be present, steel with less than approximately 0.010% sulphur was considered necessary in order to minimize the risk of lamellar tearing. As mentioned in Chapter 3, Section 3.11, there is a European Standard to control steel properties perpendicular to the surface (European Standard 10164).

Applying a low strength, notch tough weld as a buttering layer to the surface of the suspected plate or rolled steel member has been widely used. The buttering layer extends about 19mm (¾ inch) to 25mm (1 inch) beyond the toe of the weld and is about 6mm (¼ inch) to 10mm (3/8 inch) thick (see Figure 4.16b). Ultrasonic testing prior to welding, to check for non-metallic inclusions and subsequent to welding, to check for propagation of defects, can significantly assist in the quality assurance and control regarding this issue.

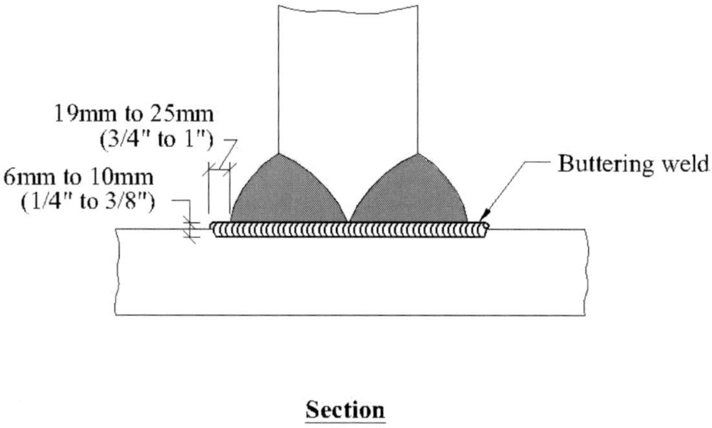

Buttering with a Low Strength, Notch Tough Weld

Figure 4.16b

Also refer to Chapter 3, Section 3.11 for other tests available. Subsequent to welding, lamellar tearing that surfaces is readily detectable by visual inspection, dye penetrant or magnetic particle testing. Ultrasonic testing is needed to detect internal cracks. It is generally considered that lamellar tearing may be more possible within sections with flanges greater than 38mm (1½ inches). However, lamellar tearing in sections with flanges less than 38mm (1½ inches) has been observed in recent years as mentioned in Chapter 3, Section 3.11(see Figures 3.15a and 3.15b). The column shown in Figure 3.15b is a W14x145 has a flange thickness of 27 mm (1-1/16 inch).

Stress Corrosion Cracking

This type of cracking is associated with the combination of stress and corrosion sometimes with other elements that promote this condition. For example, some chemical agents such as caustic alkalis, sodium and potassium carbonate provide a source of molecular hydrogen that migrates to the laminar discontinuities and induces high stresses causing cracking. This problem has occurred in pipe lines and with stainless steel in the presence of chlorides with susceptibility augmented by increase in temperature. Residual stresses from welding can increase the susceptibility for this condition to occur.

Strain-age Embrittlement

The root pass of multi-pass welds is subjected to straining and reheating by the subsequent passes. Small cracks form adjacent to the weld boundary. When reheated by successive weld passes, embrittlement occurs such that added stresses cause the crack to propagate and fracture.

Corrosion in the Vicinity of Welds

Corrosion occurs when there is a chemical reaction between a metal and its environment. Essentially it is the reverse process to the abstraction of the metal from minerals. There are various types of corrosion. Most of these involve electrochemical reactions resulting in the exchange of ions between metals due to the differences in their electro potential. A metal or area of a metal which is protected is cathodic or noble in comparison with another metal or area of metal, called anodic, which loses material.

As previously discussed, welding results in areas of steel with different properties, i.e. weld metal, heat affected zone and base metal. The metallurgical effects of welding with different cooling rates occurring in the weld metal and the thermal effects on the heat affected zone result in differences in electro potential and can be detrimental to the corrosion as well as mechanical properties of the weldments.

In media which provides high degree of electrical conductivity (e.g. sea water), weld metal and/or the heat affected zone may become an anodic area. Galvanic action then occurs such that the anodic area corrodes. Pitting can occur when a concentration cell is formed from reduction in protective passive oxide film, or solutions in contact with the material. Localized corrosion occurs when a critical value in the solution, known as the pitting potential, is reached. Welds tend to involve significant segregation which affects the microstructure with the tendency to increase the probability of attack. In cases where there is a potential for corrosion to occur, weld composition should be considered and adjusted to make its corrosion potential nobler than the base metal that is being welded. Corrosion of the base metal will then be more uniform and thus less detrimental.

A phenomena known as **Intergranular Corrosion** can occur whereby localized precipitation occurs along grain boundaries. Whole grains of metal fall away such that there is a reduction in cross section and thus strength. This type of corrosion is common with austenitic stainless steels where chromium carbide appears at the grain boundaries during heating and cooling cycles when welding. Control of this phenomenon can be achieved by keeping the welding heat to a minimum, reheating, restricting the carbon content and adding alloying elements (e.g. niobium, tantalum and titanium).

In Service Weld Cracking

The presence of incorrect profiles and/or structural discontinuities, such as slag inclusions and other nonmetallic inclusions, notches, undercut, incorrect terminations, blow holes craters, etc. can cause significant stress concentrations. See Figure 4.17 for typical defects that can occur. Dependant upon the nature of the in service stresses [e.g. high cycle fatigue such as occurs in bridges, low cycle fatigue such as occurs in earthquakes] propagation of these defects can occur. The subject of crack/defect propagation is described in more detail in Chapter 2. Furthermore, the use of inappropriate weld configurations can cause failures. For example, partial penetration butt welds can have poor fatigue resistance. Also, the fatigue strength of plug welds tends to decrease with increase size of the plug weld [Munse (1964)].

4.1.6 Effects of Fabrication Procedures

Arc Strike

This occurs when the welder accidentally strikes the electrode onto the base metal near to the intended weld. This can cause undesirable hardening of the steel in the area of the arc strike.

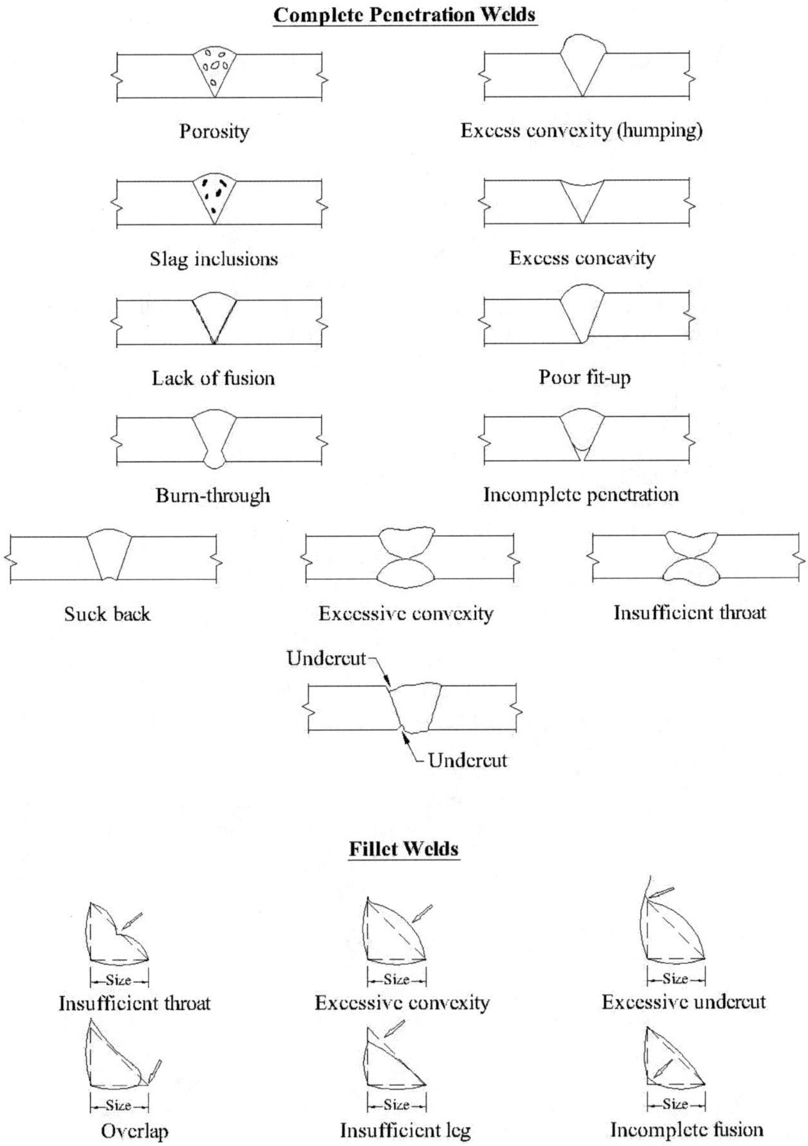

Typical Weld Defects

Figure 4.17

Weld Spatter

Weld spatter results in weld deposit being emitted away from the intended weld. This may be due to use of incorrect electrode or welding parameters such as the welding current. Again, this can cause undesirable hardening of the steel in the area of the weld spatter.

Tack Welds

Tack welds are often needed to set up assemblies of parts to be joined before the intended welds are carried out. However, since tack welds are usually small fillet welds, rapid cooling can occur which may result in low ductility and toughness in the weld metal and heat affected zone. Preheating the steel before tack welding can assist in obtaining improved ductility and toughness characteristics. Also see section 4.1.6 below with regard to intermixing of welds.

Intermixing of Welds

Intermixing of welds, which can occur due to repairs or field welds added over or adjacent to shop welds, can be a problem if the welding electrodes or welding processes are different. It has been found that, in some cases, intermixing can cause deterioration of the fracture toughness of the weld metal [Quintanna and Johnson (1997, Part I), Quintanna and Johnson (1997, Part II), and Quintanna and Johnson (1998, Part III)]. They found high dilution of Flux Core Arc Welding gas shielded (FCAW-G) from underlying FCAW-S welds significantly reducing the fracture toughness of FCAW-G. Also, there is some reduction of the fracture toughness of Shielded Metal Arc Welding (SMAW) welds when intermixed with Flux Core Arc Welding, self shielded (FCAW-S). Intermixing of welds should not be permitted without the approval and knowledge of the engineer and should be verified by testing such as the test method given in AWS D1.8 which involves determining the Charpy Vee Notch toughness of the welds.

Insufficient Allowance for Weld Shrinkage

Inadequate allowances to accommodate for distortion due to weld shrinkage. This is discussed in more depth in section 4.1.4.

Back Up Plates (Bars)

Back up plates are often used for complete penetration vee groove welds. They can be a source of defects at the root of weld often in the form of slag inclusions. Furthermore, it has been found that significant residual stresses occur immediately adjacent to the back up bar [Dong and Zhang (1998)]. These residual stresses tend to reduce the fracture toughness of the weld.

Removal of the back up plate, followed by back gouging, and depositing weld usually improves this condition by minimizing defects.

4.1.7 Repairs

Repairs of any nature have to be carried out with much care. Even grinding, machining, chipping and or gouging, for repairs to correct poor profiles, needs to be carried out such that weld metal and/or parent metal are not nicked or gouged.

In some cases, removal of welds is required. Great care needs to be taken in order not to significantly affect material properties during the removal process. Excessive heat and rapid cool down can increase hardness and reduce fracture toughness of the material. Use of preheat should be considered for the removal process. Care should be taken not to intermix welds unless it can be demonstrated that fracture toughness deterioration is not significant. The installation of repair welds tends to induce significant localized residual stresses which may be susceptible to cracking. Again, use of preheat and, in some cases, post-heat may need to be considered. Use of magnetic particle testing at the first and final bead can assist in checking for possible cracks.

Repairs to existing steel structures represents increased challenges due to presence of unknown locked in stresses and, in many cases, minimal knowledge of the properties of existing materials.

After the 1994 Northridge Earthquake, many existing steel moment frame buildings were required to be repaired in Los Angeles. Standard repair procedures are currently given in FEMA 352 (2000) which followed interim recommendations by FEMA given in FEMA 267 (1995) and FEMA 267A. In some cases, existing complete penetration welds required either partial removal or complete removal and replacement particularly at the bottom flange beam connection to the column flange. In some cases, removal, using the air arc process, resulted in the propagation of cracks as reported by Brandow and Maranian (2001) (see Figures 4.18a and 4.18b). Brandow and Maranian recommended removal commencing at the top of the bottom flange weld in order to mitigate the potential for crack propagation. Brandow and Maranian also gave recommendations for the replacement of column flanges and column webs which had cracked. Much care is needed to minimize residual stresses and maintain good fracture toughness characteristics.

A repair procedure, pioneered by the late Dr. Warner Simon, known as the weld overlay repair method, was developed and used on several steel moment buildings in Los Angeles, damaged by the Northridge Earthquake (see Figure 4.19). The method involved depositing fracture tough welds over the existing welds with minimal removal of the existing weld eliminating the potential for crack propagation. Small component static, low cycle fatigue and drop weight tests were carried out which showed very good results [Simon et al (1999)]. Large scale testing was carried out on beam to column pre-Northridge connections repaired using weld overlays which also showed good performance. The repair method is documented in Anderson et al (2000), Brandow and Maranian (2001), Maranian and Simon (2002), Simon et al (1999), and Simon et al (1988). Further testing was recommended by the researchers particularly for use in other applications.

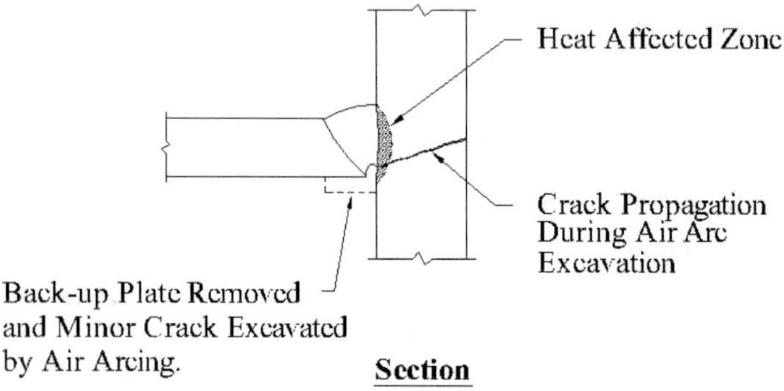

Crack Propagation into Column Flange During Minor Crack Repair

Figure 4.18a

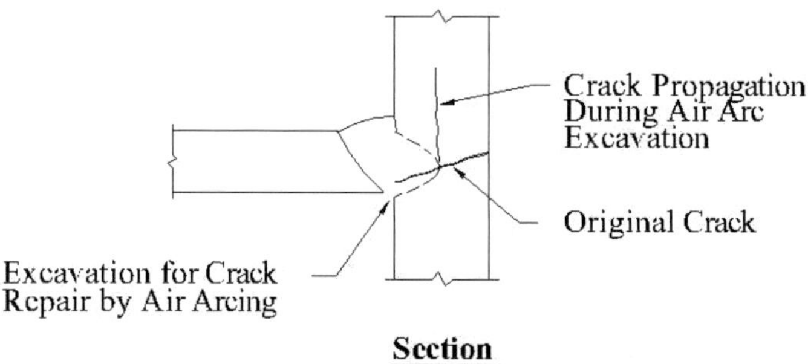

Crack Propagation into Column Flange During Column Flange Excavation

Figure 4.18b

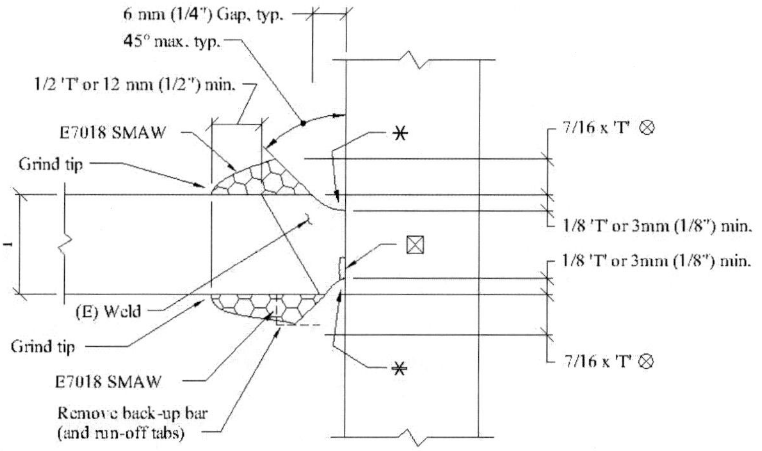

Stage 1

✳ - Grind groove in existing weld.
⊗ - Based upon 45° groove angle. Increase accordingly if smaller angle used.
⊠ - Existing minor defect may remain provided. No indication in column flange exists.

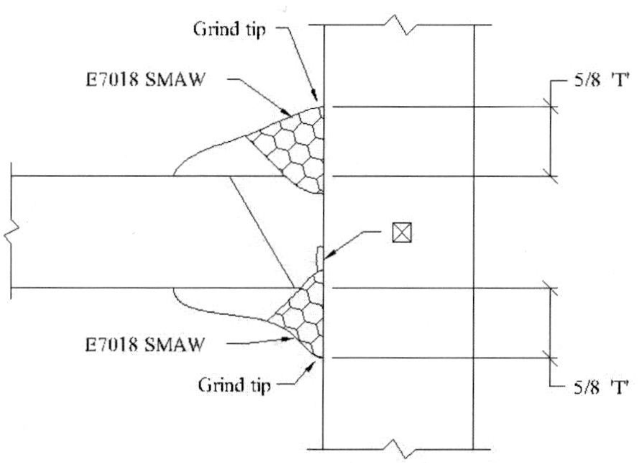

Stage 2

Weld Overlay - Class A Repair Using SMAW

Figure 4.19

4.1.8 Quality Control of Welds

Methods of evaluating welding and the quality of welds include the following:

Visual inspection before welding

This includes verifying fit up of the parts to be welded including root gap, backing bars, tack welds including preheat requirements, edge preparations, allowance for distortion, cleanliness, etc.

Visual inspection during welding

This includes verifying the correct filler metal on electrodes are used; ensuring that the equipment is satisfactory and compatible with the welding operation, verify the welder is certified for the type and position of the weld; verify preheat is applied correctly (if required), ensure the parameters (voltage, amperage, travel speed, etc.), required by the approved welding procedure specification, are followed; observe the welding operation, check for interpass temperatures, check for chipping, gouging and grinding. The inspector should also observe all repair work to verify conformance with approved repair procedures.

Visual inspection after welding

This includes checking post heat (if required), cooling rate (if required) checking weld size, conformance to profile, surface defects including cracks, undercut, surface porosity, incomplete root penetration, distortion, arc strikes, weld spatter, etc.

Dye-penetrant testing

This method, following the thorough cleaning of all dirt and film, involves the application of a liquid penetrant which seeps into any opening in the surface by capillary action. After a lapse of time, the surface is cleaned, excess penetrant is removed and a powder applied when the surface is dry. Any surface crack shows up as strongly against a white background. It should be understood that this method only detects cracks emerging at the surface. Wide cracks tend to produce a spread of the penetrant, whereas sharp cracks tend to appear as dots.

Magnetic particle testing

This method utilizes a special portable unit, involving a magnet or magnetizing coils which attach to the steel with two electrodes.

The surface is required to be thoroughly clean before magnetic powder or ink is applied to the surface. Upon application of the magnetic field, the magnetic powder or ink is attracted to the leakage of magnetic flux occurring at the defect. Again, it should be understood that this method only detects cracks emerging at the surface.

Also, abrupt changes in geometry such as sharp corners, changes in the magnetic permeability such as the heat affected zone of the weld may mistakenly indicate false defects.

Ultrasonic inspection

This method involves the use of ultrasound well above the audible range (1 to 5 MHertz) with the use of transducers (probes), in the form of a crystal, which serve as both transmitter and receiver. Essentially, the transducer, upon application of electrical voltage, sends out a beam of ultrasonic energy through the specimen. Any internal flaw reflects some of the energy and the remaining energy is reflected at the back surface of the specimen. Size and depth of defects can be determined from calibration techniques. The angle at which the beam of ultrasonic energy is transmitted can be changed in order to more readily discover defects depending on the geometry of the specimen and weld configuration.

This method is significantly dependant upon the operator's skill level. It may not always accurately determine defect size and location and it may also detect flaws which may be small and not detrimental to the material.

Radiographic testing

This method involves the use of x-rays or gamma rays applied directly through the specimen onto a radiographic film applied on the back side of the specimen. X-rays are produced when tungsten is bombarded with electrons. Gamma rays are emitted from decomposing substances such as uranium or thorium. Energy, which is not absorbed by the material, will appear dark on the film. Any areas of material where discontinuities occur, such as porosity or cracks, cause the thickness of the material to be changed. These discontinuities tend to absorb some of the energy, and appear lighter.

This method does have the advantage of having a record of test. However, it does not detect planar defects well and may not be acceptable for use on some projects due to safety concerns.

4.2 RIVETS & BOLTS

4.2.1 Rivets

Wrought iron rivets were widely used in the 19th century and early 20th century. These were used in the Titanic (1912) and as stated in Chapter 1, Section 1.1.2, significant slag in the form of stringers caused fracture of the rivets leading to parting of the hull plates and the sinking of the ship. Steel rivets, using extra soft steel, were available in the early years of the 20th century. To the best of the writer's knowledge, steel rivets are rarely used now for structural steel having been replaced by high strength bolting and welding although design criteria are still included in AISC Specifications. AISC (1930) required heating to a temperature of no more than 1065ºC (1950ºF). Typically they were heated to a cherry red complexion which is within the range of 815ºC to 982ºC (1500ºF to 1800ºF). In the past, rivets were heated in a rivet forge thrown to the riveter, a set placed against the rivet head and hammered by hand. The installation was improved by using a pneumatic or vibrating rivet gun. However, these can not be used when the temperature drops below 538ºC (1000ºF) as it tends to loosen the rivet rather than tighten it. The allowable stress, given in AISC (1930), for use of rivets installed with the pneumatic rivet gun, was higher than for other methods.

The most common sizes used diameters of 19 mm (¾ inch), 22 mm (7/8 inch) and 25 mm (1 inch) although they can be as large as 38 mm (1-1/2 inches). Holes are 2 mm (1/16 inch) larger in diameter. Rivet holes, were usually punched rather than drilled leaving significant residual stresses. These residual stresses can be alleviated by reaming a hole to a slightly larger diameter. Holes need to be accurately aligned to set them properly. Rivets have either a "button head" or are countersunk or a combination of both.

The older specification ASTM A141, used in the mid 20th century, had tensile strength of 359 MPa to 428 Mpa (52 to 62 ksi). The current specification ASTM A502 has three grades, Grades 1, 2 and 3. Typically, tensile strengths are 414 MPa (60 ksi) for Grade 1 and 552 MPa (80 ksi) for Grades 2 and 3.

The process of forging tends to make the rivet fit tighter in the hole taking up some of the slack in the hole. This is dependent upon the grip length and as the grip length is increased, clearances between the hole and the rivet tend to increase. Making the rivet fit tighter in the hole is partially offset by shrinkage due to cooling which also creates a clamping force between the surfaces. The clamping force is subject to the thickness of the parts to be joined, the rivet temperature and the driving force applied. Also, clamping force tends to increase with increase in grip and can reach yield. The clamping force is not used in the AISC Specification for allowable rivet loads since it is considered not reliable.

Driving forces increase the strength of the rivet significantly with increases of 20% for machine driving and 10% for pneumatic hammer. As expected, this is accompanied by a reduction in ductility. Tests [RCSC, (2001)] on ASTM502 rivets show little effect on the tensile strengths due to the soaking heat temperature. Tension tests [RCSC, (2001)] have also shown that tensile strength of rivets decreases with increase in grip length. Shear tests (RCSC, 2001) have shown that, although the long rivets demonstrate greater deformation than short rivets at the initial load stages, ultimate shear strength is not significantly affected.

Tests on rivets for combined tension and shear [RCSC (2001)] show, as expected, greater deformation capacity for rivets with a low shear/tension ratio compared with rivets with a high shear/tension ratio.

Considerable stress is imparted on the rivet head and at the corner of the head which creates a stress riser. Primarily as a result of the work hardening of rivets particularly at their heads, the tensile and fatigue strengths of rivets are lower than those of bolts.

It is interesting to note that, according to Tobriner (2006) steel buildings, with riveted connections, appeared to have performed well in the April 18, 1906 San Francisco Earthquake. This is also confirmed by Hamburger and Meyer (2006) in their review of the performance of steel frame buildings with infill masonry walls. Rivets were used extensively in buildings to construct built up shapes, for beam to column connections and for the connections of diagonal steel bracing. At that time, buildings were designed for lateral forces due to wind only [typically 1.44 KN/m^2 (30 pounds per square foot)]. The 1906 San Francisco Earthquake, which was a magnitude 7.9 earthquake [significantly greater than the 1989 Loma Prieta (San Francisco) and 1994 Northridge (Los Angeles) earthquakes], was also followed by an extensive fire. According to Tobriner (2006), it was difficult to pinpoint earthquake damage after the fire. However, it was thought that the damage due to the earthquake was far less than from the fire. Sheared rivets were found in the Ferry Building which had braced frames utilizing eye bars, the Flood Building and the San Francisco City Hall, which had just been completed after 20 years of construction. According to Tobriner (2006), nearly all of the rivets at splice plates connections to the columns supporting the City Hall's dome tower sheared off. This appears to have partly been due to not installing diagonal bracing per the design. It appears that the shear failure of rivets in the 1906 San Francisco Earthquake, where occurred, typically were due to substantial overload for which they were not designed for. To the best of the writer's knowledge, no failure of connections in older buildings using rivets was reported in the 1994 Northridge Earthquake.

Experimental tests by Bruneau et al were carried out on a non-retrofitted riveted stiffened seat angle beam to column connection for seismic performance [Bruneau et al (1998)]. The tests were performed on a portion of a steel frame from an existing building built in 1910 and demolished in Ottawa, Canada. The connections comprised riveted stiffened seat angle connections with top restraining angles typically used in many old buildings. The existing connections, not intended for moment capacity, showed "considerable moment capacity and exhibit a relatively ductile hysteretic behavior". However, the hysteretic curves were pinched. Slippage of the rivets, due to insufficient clamping force and lack of fit in the rivet holes contributed to the distinct pinching. Failure of the connection occurred due to shear failure of a rivet in the seat angle after several cycles.

Fisher et al (1998) describe tests developing fatigue life of rivet connections and were able to develop an S/N curve and a fatigue limit. Factors affecting the fatigue life of riveted connections include the clamping force, the bearing condition, and the formation of the hole (drilled, punched, and reamed). Experimental fatigue testing on riveted shear splices indicated cracking in the connected material and not the rivet itself.

4.2.2 Bolts

History and Development

During the early decades of the 20th century, finished bolts were permitted only for small structures, secondary members, purlins, girts and doors (AISC, 1930). Bolts were also used for erection purposes.

Research carried out by Dr. C. Batho at Birmingham University in England, during the early 1930s, led to the use of high tensile bolts with controlled torque [Steel Designer's Manual, 4th Edition 1972; RCSC, (2001)]. Bolts with minimum yield strength of 373 Mpa (54 ksi) could be tightened to give an adequate margin of safety against slippage of the connected parts. The work by Batho solved the problems of developing a practical design method for highly redundant structures and for substituting high strength bolts for rivets. However, the British industry made no use of this pioneer work. In 1947, the Research Council on Riveted and Bolted Structural Joints was formed in the USA. Studies on high strength bolts and rivets were carried out and a tentative specification for use of high strength bolts was first approved in 1949.

The first specification was issued in January 1951. Further studies in the early 1950s were carried out on installation procedures, slip resistance and repeated loadings. Ironically, the British Industry had to pay the American licensors, who had patented the devices (Steel Designer's Manual, 4th Edition, 1972). The British eventually issued British Standards BS 3139 on bolt material in 1959 and BS 3294 on design procedures in 1960.

Further research led to increase in the allowable stresses, modification of bolt tightening procedures, the use of higher strength A490 bolts, galvanized bolts and slotted holes.

Three types of bolts are commonly used in the U.S.A. That is ASTM A307 Grade A carbon steel bolt, ASTM A325 high-strength steel bolt and ASTM 490 quenched and tempered alloy steel.

When high strength bolts are pretensioned they may be regarded as pretensioned or slip critical. Pretensioned bolts are specified for load conditions including significant load reversal, joints subject to fatigue and connections subject to seismic demands. Slip critical connections, which depend on special surface preparation at the faying surfaces, are used for conditions including fatigue with load reversal, oversized and slotted holes and where slip is detrimental to the performance of the structure.

Specified minimum tension (70% of ultimate strength) for high strength bolts, when first introduced, were carried out by the calibrated wrench method using torque control. A washer is required under the turned element. This method can be subject to significant variation due to friction between the nut and the bolt threads and the nut and the washer. Even though lubricants are used, they can become contaminated with dirt and affected by moisture or rain. Exposure of the steel to weather causes increases in the friction which reduces the tension for the same torque applied.

The turn of the nut method was introduced into the specification in 1960 for A325 bolts and involves a specified additional turn from the snug tight position. The use of the turn of the nut method for A490 bolts was introduced in 1964. More testing and calibration led to further modifications in 1974.

Other methods of obtaining the specified tension in the bolts are by using load indicating washers, tension control bolts and by calibrated wrench pretensioning. Load indicating washers involve special washers with projections which compress closing the gap between the nut or bolt head and the flat portion of the washer. Tension control bolts incorporate a spline extension with a circular notch. The notch is calibrated to fail in shear at a torque that corresponds to the required tension. Washers are required under the nuts. Calibrated wrench pretensioning involves the use of a torque wrench which is calibrated for the required torque. Since there can be significant variation in the pretension due to several factors including surface conditions, torque, wrench power supply, etc., verification of the calibrated torque wrench needs to be carried out daily.

Use of high strength bolts prior to 1985 required that they be tensioned. Following practices developed in Europe, use of non-preloaded high strength bolts was then allowed where slip critical connections are not required. This permitted higher strength bolts to be tightened only to a snug-tight condition allowing slip to occur and shear to be taken in bearing.

Installation

The installation of bolts is very important and can significantly affect bolt performance. Issues include the following:

- Attaining required tension for slip critical and pretensioned bolts. For example, not following the correct procedures for load indicating washers such as installing the special load indicating washers the wrong way. Exposed steel can affect calibrated torque wrench and turn of the nut methods. Use of oil on surfaces, which reduces slip resistance, can assist in attaining torque values. Improper installation of washers and nuts in tension control bolts resulting in bolt thread stripping has been reported.

- Slip critical and pretensioned. Sequencing tensioning of individual high strength bolts such that the prestress in previously tensioned bolts are relaxed thus reducing the total clamping force.

- Inadequate certification and verification of the specified bolts and nuts.

- Improper storage of bolts allowing dirt and moisture to occur on the bolts.

- Mixing A325 and A490 bolts. Not installing 'x' bolts (threads excluded from the shear plane and which have higher load capacities than bolts with threads in the shear plane) where specified.

- Ignoring requirements to carry out pre-installation testing on samples to verify correct assemblies, pretensioning method, procedures, inspection as well as identifying potential problems.

- Lack of fit of holes. Sometimes excessive reaming is carried out to make it fit. However, the AISC Specification only permits an increase in hole diameter of 0.8 mm (1/32 inch). Excessive reaming can cause uneven bearing of bolts. Misaligned bolts result in less movement and increase joint stiffness. However, with increases in yield and tensile strength, less ductility is available for distributing the load which may result in failure of the most high loaded bolt(s).

- Bolts not installed straight such that additional stress is imposed on one side of the bolt head and nut. AISC Specifications require sloped surfaces, in contact with the bolt head or nut, not to exceed 1:20. Hardened beveled washers are required if this limitation is exceeded.

- Weld spatter on the faying surfaces (surfaces of each element to be joined). Weld spatter can cause undesirable gaps between the faying surfaces as well as hardening the surfaces of steel.

- Gaps between the plies causing undesirable secondary stresses in the connection. Plies need to be pulled together so that all plies are in firm contact. This is usually achieved by tightening from the most rigid to the free edges. Where thicker members occur, tight contact across the whole surface may not always be possible. This may not necessarily be detrimental if the specified bolt tension provided such that the total clamping force is maintained.

- Lack of washers where required. Hardened washers (ASTM F436) are required for certain conditions including where torqued by the calibrated wrench method, steel material less than 276 MPa (40 ksi) for ASTM 490, at oversized and at short slotted holes (Ref ASTM Spec for A325 or A490). 8 mm (5/16 inch) washer plates are required at long slotted holes for all A325 bolts and A490 bolts 25 mm (1 inch) diameter or less. Washers are required to prevent galling of the connected material and a more even distribution of the bolt clamping force.

- Incorrect nuts. A490 bolts require a higher grade nut in order to minimize the potential for failure of the nut or bolt threads by stripping. Failure by tension in the bolt threads is preferred to stripping of the threads since it is more easily detectable.

- Faying surfaces at slip critical bolts having loose corrosion, oil or dirt affecting the friction coefficient. Tight mill scale is acceptable.

- Faying surfaces that have incorrectly been shop primed painted at connections using slip critical bolts. Note, paint on the faying surface may be used at slip critical connections provided it is tested to determine the slip coefficient for use in the design.

- Use of excessive filler plates in bearing type bolted connections. Loose filler plates up to 6 mm (¼ inch) thick are acceptable, whereas 19 mm (¾ inch) loose plate can result in a reduction in bolt capacity of about 15% primarily due to bending of the bolts. It should be noted that loose filler plates are acceptable when using slip critical bolts, which are not expected to go into bearing, since resistance between the plies is provided by friction with no reliance on bolt bending.

- Incorrect use of oversized holes or long slotted holes in connections where slip critical resistance can be exceeded. Substantial displacements are likely to occur. It should be noted that frictional resistance for oversized and slotted holes tends to be less than for standard holes.

Figures 4.20a and 4.20b address conditions discussed above. Figure 4.20c shows a field condition where significant lack of fit occurred.

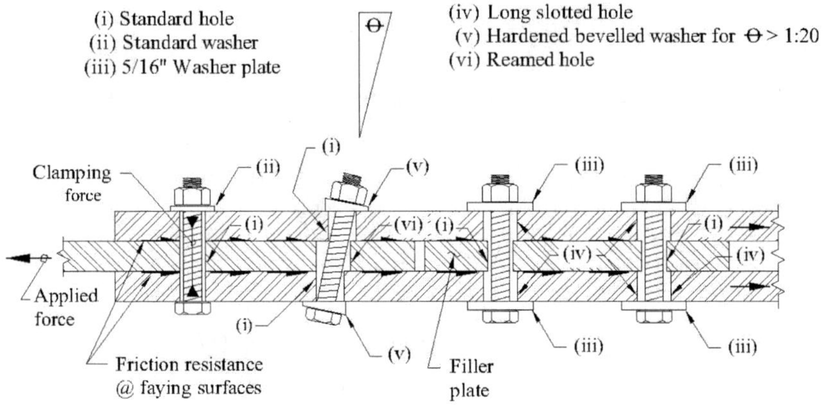

Spliced Slip Critical Bolted Connection – A325 bolt

Figure 4.20a

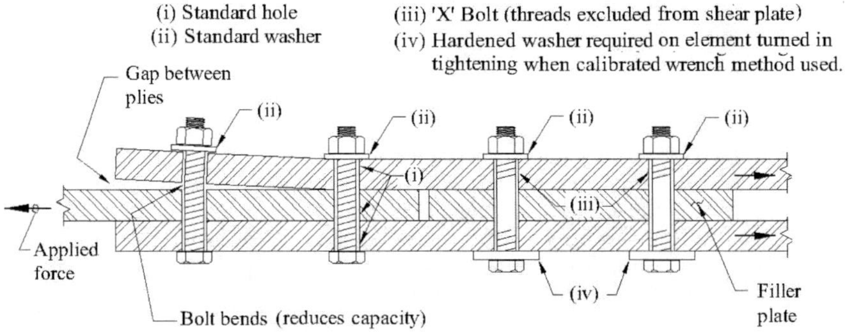

Spliced Bearing Bolt Connection – A325 bolt

Figure 4.20b

Poor Bolt Fit-up

Figure 4.20c

Other Possible Problems with Bolts

Hydrogen embrittlement (see Chapter 4, Section 4.1.5) can also occur with bolts particularly where atmospheric exposure conditions are such as to encourage it. The atomic hydrogen, which tends to migrate towards areas of high tensile stress, produces a hard martensite which is susceptible to cracking. Stress corrosion can also occur under similar conditions and corrosive action tends to occur along areas subjected to tensile stress. Both hydrogen embrittlement and stress corrosion can result in delayed brittle failures and tests have shown that higher strength steels tend to be more sensitive to these effects. The oil, gas and petroleum industries have been particularly affected by these problems due to the more onerous atmospheric conditions that prevail.

Use of hot-dip galvanized bolts can provide good protection. However, improper galvanizing or damage, such as to cause breaks in the zinc film, can promote the entry of atomic hydrogen into the metal and/or promote corrosive action. Tests showed a higher susceptibility to hydrogen embrittlement and stress corrosion for ASTM A490 bolts and these tests led to AISC prohibiting these bolts to be galvanized. Galvanizing of bolts can cause difficulty in attaining bolt tensions and hardened nuts (same as for ASTM A490 bolts) are required to prevent thread stripping. Relaxation of the pretension in galvanized bolts can be particularly significant particularly in joints comprising several plies coated with galvanizing.

Bolt failures were found following the 1994 Northridge Earthquake. Failures included complete shear failures at concrete tilt up wall shear plates (see Chapter 1, Section 1.1.9), shear failure of bolts at moment connections, and tension failures at seated connections at the end of moment frame girders.

An excellent discussion on the phenomenon of "fretting fatigue" is given in Fisher et al (1998) which primarily address high cycle fatigue in steel bridges. Where bolts are pre-loaded, load transfer occurs by friction until loads applied exceed frictional resistance and the bolts then go into bearing. Minute slippage can cause the initiation and propagation of cracks at the extremities of the joint. For non pre-load bolts, high tensile stresses in the connected parts adjacent to the hole can occur. This can result in fatigue cracks at the edge of the hole or the barrel of the hole, propagating across the net section of the connected part. Fisher et al (1998) state that "fretting fatigue" is highly unlikely with non pre-load bolts but that the possibility of failure from either mode should be checked for pre-loaded bolts.

Bolt failures from an overhead traveling crane have been encountered according to Patel (2008). The bolts were not overloaded. However, failure occurring at the thread root located at the face of the nut was identified due to insufficient torquing of the bolts.

With regard to "threaded rods" used for base plates, Fisher et al (1998) advise that typically they have poor fatigue resistance. Also, the fatigue resistance is not significantly affected by steel grade, thread size, thread forming and rod diameter. This is of particular concern for structures such as sign structures where a large number of cycles at low stress due to wind can occur.

4.3 FABRICATION

4.3.1 Mechanical Cutting Processes

Drilling

Drilling, utilizing twist drills, is sometimes augmented with a tube in each flute to allow for oil to lubricate the surfaces and enable the chips to be washed away. In fabrication shops, the drills are electrically operated or air-operated hand drilling machines. Drills include the radial arm drilling machine, which has arms capable of turning in a full circle and also vertically, bench drills or upright drills. Air (pneumatic) operated drills, which have the advantage over electric drills of not overloading, are often used for reaming holes.

Magnetic or "limpet" drills are used for existing steel. These utilize an electro-magnetic base which, when switched on, attaches the drill to the work. The drills for this type of work are electric or pneumatic usually when a large number of holes are required to be drilled.

Manual drilling of holes tends to be expensive since it requires specialist skills for marking, etc. Automatically operated machines are available that are capable of drilling multiple holes without marking and are able to compensate for rolling tolerances. The process of drilling may give some reduction in the fatigue life. However, the reduction in fatigue life due to drilling is usually less than that of holes installed by punching. The use of preloaded bolts, inducing high local compressive stresses, tends to minimize the reduction in fatigue life due to drilling holes.

Guillotining

Guillotining is carried out by machines, operated mechanically or hydraulically, with a movable shear blade and a fixed bottom blade on flat plates. The movable shear blade penetrates approximately 30% to 40% of the material from the top edge and the remainder is separated by work hardening and shear tearing action. The angle of the movable shear blade is adjusted to suit the hardness of the steel (see Figure 4.21). The softer the steel, the higher

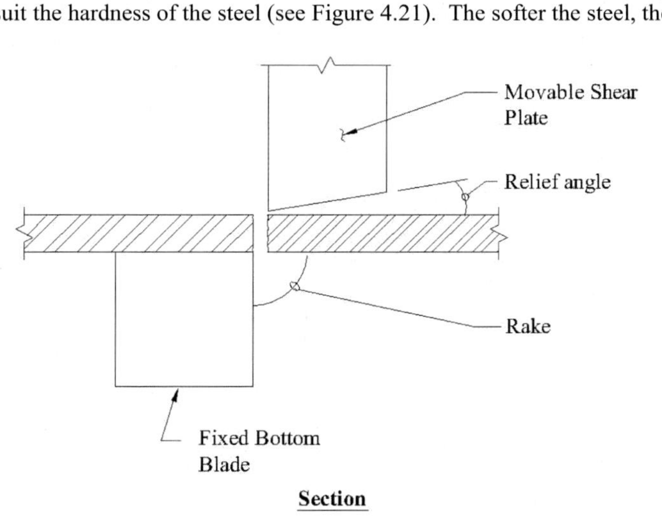

Guillotine Cutting Process

Figure 4.21

the rake angle. High strength steels and stainless steels, which have high shear strength, require greater force to cut than an equal thickness of low carbon steel. This method cannot be used on hard materials (more than Rockwell B60). Remaining surfaces, following guillotining, may contain work hardened steel.

Cropping

Cropping is similar to guillotining except that sections can also be cut as well as flat plates. Clamping of the piece is achieved by hydraulic or pneumatic means and the process can be automated with feed conveyors.

Punching

Punching, which is similar to guillotining, operated mechanically or hydraulically, can be used to make circular, slot and other shaped holes in plates and sections. The plate or section is punched from the top of the hole causing fracture to occur through the remaining section. Greater force is required than with guillotining since the punch or die applies its force to a circumference rather than a straight edge. The punch is usually flat with a nipple at the center for alignment. The punch is hydraulically driven and the process can be fully automatic on a conveyor system. Several holes can be punched at the same time through sections including angles, channels beams and columns.

The effects of punching can tend to cause a "bell" like shape and can result in imperfections and work hardening around the edges. In some cases, this can cause edge cracking. Since stress concentrations occur around the holes, this process may not be suitable for some structures including those subjected to fatigue.

Sawing

Types of saws include the circular saw, hack sawing machines, band saws and abrasive cut off wheels. Saw blades are usually lubricated with an oil emulsion pumped in a jet or sprayed onto the blade and the material. The cooling effect allows for high cutting speeds.

Abrasive cut off, or friction disc sawing utilizes aluminum oxide or silicon carbide. It is an effective cutting process for high strength and/or hard steels usually providing acceptable finishes that do not require milling.

Abrasive Water-Jet

In this process, high pressure intensifier pumps inject water at pressures between 207 MPa to 380 MPa (30,000 p.s.i. to 55,000 psi) through a sapphire stone nozzle and then mixed in a chamber with an abrasive, usually Garnet sand or aluminum oxide. The mixed stream then leaves the chamber at a pressure between 214 MPa to 255 MPa (31,000 psi to 37,000 psi) and velocities of 760 m (2,500 feet) per second. The water and abrasive material are then collected in a catcher.

This process offers many advantages compared to contact cutting and thermal cutting processes. There is no dust or smoke but noise levels can be high (90 dB to 110 dB or higher). High precision edges can be made. It is capable of cutting steel including beams without causing any heat affected zone and thus no change to the microstructure. The process has also been used to remove rivets, less than 50 mm (2 inches) in diameter, from existing steel structures.

The initial cost of machinery is high and mixing nozzles have to be replaced regularly depending on the type of abrasive used.

4.3.2 Thermal Cutting Processes

Oxygen Gas Cutting

The process, which is one of several oxygen cutting processes, involves the generation of heat from oxygen being ignited by acetylene. The heat of combustion provides preheating of the steel at about 675°C (1250°F) and then causing a chemical reaction which is essentially rapid oxidation of the steel by introducing a stream of oxygen. The resulting iron oxide melts and is removed by the stream of oxygen. Different types of tips (nozzles) and sizes of tips are available to suit different applications such as thickness of material to be cut, removing weld defects, and edge preparation.

The flame cut tends to cause a narrow zone of hardness dependent upon the carbon and alloy content. The hardness can be reduced by preheating to about 540°C (1000°F) and/or removed by grinding or machining. The measures may be necessary to prevent cracking particularly at stress concentrations. Careful control of the levels of oxygen and acetylene are necessary to ensure too large a flame does not cause the top edge to be melted during the preheating stage. Insufficient or an excess of either oxygen and/or acetylene can cause detrimental effects to the steel. The cutting speed also needs to be carefully controlled as poor control can affect the profile of the cut and cause irregularity. Furthermore, preheat levels need to be carefully controlled. Preheat too small results in uneven surface whereas excessive preheat can result in excessive melting. Cut edges usually require grinding.

Oxygen Lance Cutting

This process, which is another oxygen cutting process, involves a lance, in the form of a pipe or tube (approximately 3 mm to 6 mm (1/8 inch to ¼ inch) in diameter) which carries oxygen to the point of cutting. The edge of the piece is first preheated with a welding torch. Similar to the Oxygen Gas Cutting process, rapid oxidization occurs, with the lance, in supplying iron to the chemical reaction, being consumed in the pool. This process is capable of cutting large steel or cast iron sections.

Arc and Plasma Cutting

The principles of these processes are to provide sufficient cutting energy by heating gas by producing an electric arc between electrode and the work piece. The arc creates ionized gas which is called plasma. Plasma, neither gas liquid or solid, may be considered the fourth state of matter. As the plasma passes through a small orifice, it accelerates and becomes more ionized producing temperatures between about 8,900ºC (16,000ºF) and 13,300ºC (24,000ºF). The resulting high velocity jet concentrates on a small area and the metal is immediately melted and blown away.

In the **Oxygen Arc Cutting** process, oxygen is introduced at high pressure into the plasma stream below the orifice to cause the steel to oxidize and remove material from the cut. The electrode is usually made of steel with internal aluminum or magnesium wires. This process is capable of cutting material up to 75 mm (3 inches thick). The process tends to be used for salvage and repair work. Special care needs to be taken due to the high heat and smoke produced.

The **Air Carbon Arc Cutting** process utilizes, in parallel, a carbon electrode and a jet of air, to remove the molten metal. Since the metal is melted and removed quickly, the surrounding areas do not reach high temperatures resulting in metallurgical benefits reducing the possibility of cracking and/or distortion. It is typically used manually to remove defective welds and metal, root gouging and preparing grooves for welding. The process causes a high noise level and the large molten metal removed can become a fire hazard.

The **Plasma Arc Cutting Process** uses the same principles as the Oxygen Cutting processes except that a high velocity jet of ionized gas, which is oxygen, is used to remove molten metal (see Figure 4.22). The plasma gas and ionized gas are issued through a nozzle. The gas used is air when used with low current (manual) systems and nitrogen or argon or hydrogen when used with high current (automated) systems. A variation of this process is where water is injected to control the arc construction and also to cool the metal surface when ejected from the nozzle in a conical spray. The cut edge produced is usually as good as or better than that produced by the Oxygen Cutting processes. Also, the Plasma Arc Cutting Process has less of a detrimental affect on base metal than the Oxygen Cutting Processes. Distortion and width of heat affected zone are usually small.

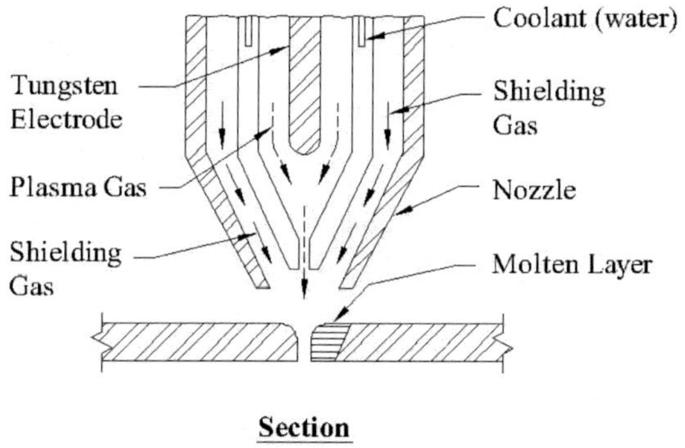

Plasma Arc Cutting

Figure 4.22

4.3.3 **Folding, Bending and Cambering**

Steel plate, up to one inch, can be bent by cold forming using press brakes which are operated mechanically or hydraulically. Dies are used in order for the press brake to form the bend (see Figure 4.23). The die width needs to be wide enough to avoid fracturing. For a 90º bend and a 10 mm (3/8 inch) mild steel plate, 8 times the plate thickness usually is normally adequate. This ratio needs to be increased for thicker plates and high tensile steels.

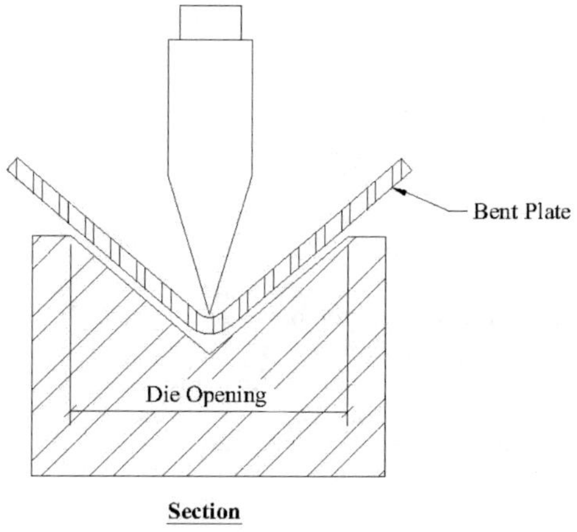

Press Brake

Figure 4.23

Cambering of wide flange beams is mostly carried out by placing the wide flange beam in a press with hydraulic rams cold bending the member at the third points. Generally, the forces required to camber members are relatively small. However, pre-drilled holes may encourage localized yielding in that area which could cause fracture when cambering is applied.

It should be recognized that the forming process, during curving of members, utilizes a significant proportion of the strain capacity of the steel such that strain demands can cause local buckling particularly web buckling of wide flange members. Furthermore, ductility demands due to in service loads may need to be limited.

An excellent discussion on cold bending of wide flange members is given in Bjorhovde (2006).

4.3.4 Heating of Steel for Straightening/Cambering

Straightening or cambering steel is achieved by steel being stretched or compressed beyond yield and retaining the deformation. By heating the steel to a temperature between 370º and 700ºC (700ºF and 1300ºF), the yield value is significantly reduced. An excellent discussion on these aspects is given by Avent and Mukai (2001).

An application of heating of steel is in cambering of beams which is usually only carried out when the size of the beam exceeds the capacity of the cold-camber process (see Section 4.3.3). Heat is applied in a vee shape starting with the apex and being advanced progressively from top to bottom. Initially, the top of the member moves up. As more heat is applied, moving progressively down the apex, the expansion due to heat moves the member down. After completion of heating, the steel cools and contracts greater than due to expansion by heating, due to contraction occurring uniformly compared with expansion which occurred progressively. These result in a net deformation occurring at the end of cooling (see Figure 4.24)

Similar approaches using heating are carried out in straightening of distorted members. According to Avent and Mukai (2001), testing has shown that yield stress in the heated regions can increase by approximately 10 percent, modulus of elasticity reductions of 8 to 23 percent have been found and there can be as much as one third reduction in ductility. Apparently, according to Avent and Mukai (2001), changes in notch toughness and hardness are not significant except that small reductions were noted in quenched and tempered steels.

Also, according to Avent and Mukai (2001), "although data is sparse, there is no indication that carbon steels will have shortened fatigue life after heat straightening."

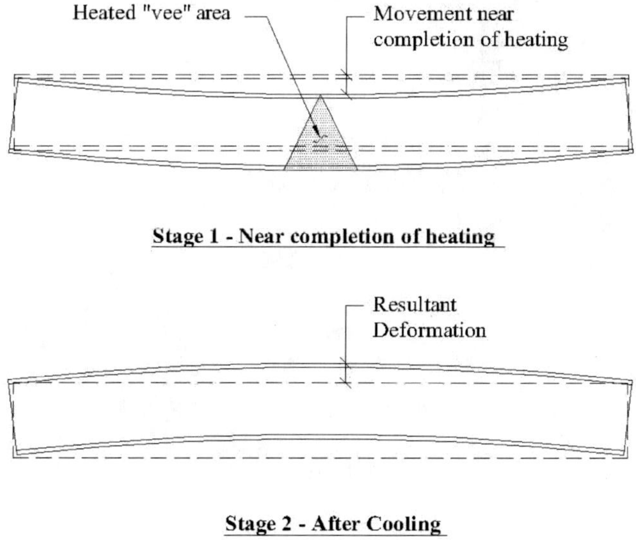

Heat Used for Straightening

[Reproduced from Avent and Mukai (2001) Copyright American Institute of Steel Construction, Inc. Reprinted with permission. All rights reserved.]

Figure 4.24

According to Avent and Mukai (2001), residual stresses due to heat straightening can be large. Studies found residual stresses of 45% of yield in plates and approaching yield in rolled sections. The high residential stresses could have implications on stability considerations including local buckling.

4.3.5 Fabrication Issues

There are requirements in codes (e.g. AISC (2005), AISC Seismic (2005), AWS D1.1) for access holes and copes to be radiused. This is important to reduce stress concentrations. Poor practices of improper access holes and copes including gouges can significantly affect performance (see Figure 4.25 for examples of good and bad copes).

Recognition of the importance of proper access holes and copes was made after the Liberty Ship failures during World War II (see Chapter 1, Section 1.2.2). Much research work was carried out on semi-circular holes and copes on welded beams following the Liberty Ship failures and is documented by Grover (1954). Tests were carried out at varying temperatures on drilled holes and flame cut holes. As expected, brittle failure occurred at

Good Cope (not a block)

Bad Cope (note notch)

Examples of Copes
[Photos courtesy of Daniel G. Luna]

Figure 4.25

temperatures of -34°C (-30°F) and below. The surface condition, at the edge of the hole, was found by researchers at that time to be an important factor. They discovered that by drilling holes followed by careful filing the edge to a smooth surface, led to no cracks even at temperatures as low as -82°C (-115°F). The

investigators carried out tests on beams with pressed notches 2.5 mm (0.1 inch) in depth in holes. Brittle failure occurred in all beams when tested at -34°C (-30°F) and -51°C (-60°F). A beam made from semi-killed steel (failed in a brittle manner at -1°C (+30°F) whereas a beam made from fully-killed steel and tested at the same temperature failed by buckling.

The American Institute of Steel Construction, following incidents of cracking in 1989, during erection and fabrication of so called "jumbo" (heavy) sections, developed requirements for larger, more elongated access holes for these members. Apparently the cracks, initiating at the access hole and in the absence of external loads, were brittle in nature and propagated through the complete section [Miller (1993)]. The "jumbo" sections tend to have a cast core within the middle of the section with a greater concentration of carbon and other alloys in this region. This can result in the cast core having poor fracture toughness. Access holes, required for complete penetration welds, cut through this region with poor fracture toughness. This led to the development of more elongated access holes which helps to reduce residual stresses developed by weld shrinkage.

For additional discussion on access holes pertaining to weld shrinkage, see Section 4.1.4.

Some additional fabrication issues include the following:

- Use of a cutting process that causes undesirable hardness which may cause cracking.

- Cut surfaces that are irregular. Grinding of the surfaces to a specified roughness should be considered where significant stress concentrations can occur.

- Consideration of cold form straightening affecting ductility.

- Consideration of heat straightening affecting ductility and causing significant and undesirable residual stresses.

RECOMMENDED READING – CHAPTER 4

Althouse, A.D., Turnquist, C.H., Bowditch, W.A., Bowditch, K.E. (2000); Modern Welding; The Goodheart-William Company, Inc.

AISC (1973); Commentary on Highly Restrained Connections Journal; American Institute of Steel Construction Journal, Third Quarter.

Avent, R.R. and Mukai, D.J., (2001); What You Should Know About Heat Straightening Repair of Damaged Steel, The American Institute of Steel Construction Engineering Journal, First Quarter, 2001.

Blodgett, Funderburk and Miller; Fabricator's and Erector's Guide to Welded Steel Construction by James F. Lincoln Arc Welding Foundation.

Campus F (1952); Effects of Residual Stresses on the Behavior of Structures; F. Campus.

Collins, J.L., (2008); The Hole Story. Modern Steel Construction, The American Institute of Steel Construction, June Edition.

Davies, B.J. and Crawley, E.J., Structural Steelwork Fabrication Volume One, The British Constructional Association Ltd. Bournehall Press Ltd.

Farrar, et al, (1975); Investigations into Lamellar Tearing: A Compendium Report From a Sponsored Research Programme Phase I-IV and Phase V-VII; The Welding Institute, England.

FEMA 288, March 1997; Background Reports: Metallurgy, Fracture Mechanics, Welding, Moment Connections and Frame Systems Behavior

FEMA 353, July 2000; Recommended Specifications and Quality Assurance Guidelines for Steel Moment Frame Construction for Seismic Applications

Fisher, J.W., Kulak, G.L., Smith, I.F.C., (1998); A Fatigue Primer for Structural Engineers; National Steel Bridge Alliance.

Funderburk, (2000); Taking your Weld's Temperature; AISC Modern Steel Construction, February.

Jones, D.A., (1996); Principals and Prevention of Corrosion, 2nd Edition, Pub. Prentice Hall, Upper Sadle River, NJ 07458.

Lancaster, J., (1992); Handbook of Structural Welding; McGraw Hill.

Lancaster, J., (1997); Handbook of Structural Welding, Woodhead Publishing Ltd., (UK).

Masubuchi, K., (1980); Analysis of Welded Structures, MIT, Pergamon Press.

Modern Welding Technology, Howard B. Cary, pub. Prentice Hall. 6th Edition, 2004
Modern Welding by Althouse, A.D,, Turnquist, C.H,, Bowditch, W.A., Bowditch, K.E., Pub. The Goodheart-Wilcox Company Inc.

Kulak, G.L., Fisher, J.W., Struik, J.H.A., RCSC (2001); Guide to Design Criteria for Bolted and Riveted Joints; American Institute of Steel Construction.

Tobriner, S., (2006); An EERI Reconnaissance Report: Damage to San Francisco in the 1906 Earthquake – A Centennial Perspective; Earthquake Spectra, Special Issue II, Vol. 22, April; Published by the Earthquake Engineering Research Institute.

Tsai, C., Kim, D., Jaeger, J., Shim, Y., Feng, Z. and Papritan, J.; Design Analysis for Welding of Heavy W Shapes." The Welding Journal February, 2001.

CHAPTER 5

Discussions and Recommendations

5.1 DISCUSSION

5.1.1 Lessons to be learned from past failures

Use of steel structures since the 19th Century, has generally been highly successful in constructing safe structures for the benefit of society. The development of steel connections, first with rivets, later by welding and the use of high strength bolts, has enabled structures to be built to accommodate the significant demands of gravity, the forces of nature, accidents, fires and terrorist acts.

Over the course of the history of steel, there have been several failures of which only a small portion has been discussed in this document. Engineers and metallurgists have gained much understanding from these failures during the history of steel. The fracture of a steel girder during fabrication and the failures of welded steel bridges in Belgium and Germany, prior to World War II, helped gain understanding of residual stresses. The failure of the Liberty Ships during World War II and the failure of other ships, shortly after World War II, led to the further development of Fracture Mechanics, first established by Griffith in the 1920s and the better understanding of the effects of low temperature, constraint, strain rate, fatigue along with fabrication issues, welding issues and material properties. This increased understanding significantly helped to reduce failures in industries such as the Ship Building Industry. However, failures, from time to time have still occurred. Not all the knowledge gained from the research carried out on Fracture Mechanics and Metallurgy has been absorbed into some industries utilizing steel. This is particularly true of the Building Industry where most engineers have only a limited knowledge of these subjects and Fracture Mechanics is hardly mentioned in codes. It seems that the lessons learnt from the failures have not always been fully passed on to subsequent generations resulting in the repetition of failures. It is the opinion of the writer that steps should be taken, by the educational institutions, to improve the knowledge of engineers in the subject of Fracture Mechanics and Metallurgy. Understanding of these subjects is important in establishing an awareness in recognizing the vulnerability of design details, components and materials to the possibility of failures not predicted by conventional static and dynamic analysis procedures By doing so, engineers can better assess the applications and limitations of materials, design and fabrication practices in order to reduce the risk of failures.

5.1.2 Size Effects and Constraint

Size effects and constraint are not, in the opinion of the author fully appreciated by engineers and adequately addressed by building codes and are thus cited in the discussion as significant concerns. Concern for size effects was expressed by Parker (1957) in the 1950s. Parker demonstrated a tendency for an increase in transition temperature with increase in thickness. Parker also noted thick plates [e.g. 25mm (1 inch)] behaving in a brittle manner at moderate temperatures whereas thin plates [e.g. 6mm (¼ inch)] behaved in a ductile manner at low temperatures. This is also noted by Eduardo Torroja in his book "Philosophy of Structures" [Torroja (1958)] with his statements "permanent residual stresses, which in very large shapes, are sometimes of such intensity that spontaneous failures occur" and "it is extremely important to know that any tri axial state of stress is dangerous and can cause failures of a brittle type". Size effect was discussed by Munse (1964) where it was stated that "the fatigue strength of a larger member might be expected to be somewhat lower than that of a smaller member because of the greater opportunity for flaws to exist than with the larger member." Burdekin [1999], also explained that the effects of increased size tend to have greater potential for failures due to fracture, fatigue and corrosion. Concern for size effects for certain modes of failure was also expressed by Krawinkler et al (1983) (see later for further discussion on Krawinkler et al's document). This is due to larger members, in comparison with smaller members, having material being rolled less, joints being more constrained, greater residual stresses due to welding and higher strains for similar inelastic rotational demands. Thus, increase in fracture toughness levels may be required for larger and heavier members. It should be noted that procedures to account for different thicknesses for weld fracture toughness is given in British Standard 7910. Burdekin and Suman (1998) used this approach to evaluate CVN requirements for different geometries of weld overlay repair details. The need to have higher fracture toughness levels for larger and heavier members is not significantly reflected in current standards in the building industry in the United States in the opinion of the writer.

5.1.3 Lack of provisions in codes to address Fracture Mechanics issues

Current building codes, for the most part, address the issues associated with brittle failure primarily by quality assurance and quality control requirements (for example, CVN weld and parent metal requirements, non destructive testing requirements). However, although, for example, in seismic design, steel is often required to tolerate significant yielding, there are no provisions to check tri axial stresses, determine stress and/or strain concentrations and account for defects other than by testing (e.g. steel moment frame connections, buckling restrained braced frames). The structural engineer is left to the use of standard practice and may overlook circumstances where potential for brittle failure can occur. Current demands on the structural engineering profession are such as to impose significant restraint on design in order to produce economic structures by only complying with the minimum requirements of the code. However, structural engineers should be encouraged to go beyond mere compliance with code and at least examine potentially problematic details.

5.1.4 Inclusion of Fracture Mechanics in structural engineering curriculum

It is the opinion of the author that at least a primer course in fracture mechanics should be included in the education of structural engineers. This may be associated with courses in metallurgy. This is considered necessary in order to give basic understanding of issues including constraint, stress and strain concentrations, fracture toughness, fatigue, residual stresses, size effects, brittle to ductile changes due to temperature variations, etc.

5.1.5 Further research

The most significant research and development program that has taken place in the USA, associated with the structural steel industry, was the SAC project which followed the January 17, 1994 Northridge Earthquake and was primarily associated with steel moment frame connections (see Chapter 1, Section 1.2.11) and which culminated in 2000. Much good was carried out in the SAC project that led to the publication of several documents including FEMA 350 (2000) on moment frames. The SAC document FEMA 355 D, State of Art Report on (steel moment frames) Connection Performance outlined the following issues as unresolved requiring "additional research to develop fully rational design guidelines"

- Reliability of details with minimum testing, in particular

  Free Flange and Weld Overlay details.
- Liberalized lateral bracing requirements for girders.
- Effects of panel zone yielding on connection performance.
- Yield mechanisms and failure modes of bolted connections.

A number of findings and issues were found to remain according to SEAOC (2002) which provides Commentary and Recommendations on FEMA 350 by the Structural Engineers Association of California, Seismology Committee. Recommendations for further research on Steel Moment Frames for Seismic

Lateral Resisting Systems are also given in SEAOC (2002). Twelve topics of further research were noted. Of these five were considered to be of highest priority as follows:

- As constructed weld interface
- Additional connection tests
- Panel zones
- Low cycle fatigue
- Deep columns

There has been some ongoing research and testing on Steel Moment Frames for seismic design since the culmination of the SAC project in 2000. The research and testing has included additional connection tests and studies regarding continuity plates, panel zones, low cycle fatigue and deep columns.

It is also the opinion of the author that at least some beam to column tests should have also applied horizontal loads to cause column yielding since this is predicted by analysis for earthquake motions (see Figure 5.1).

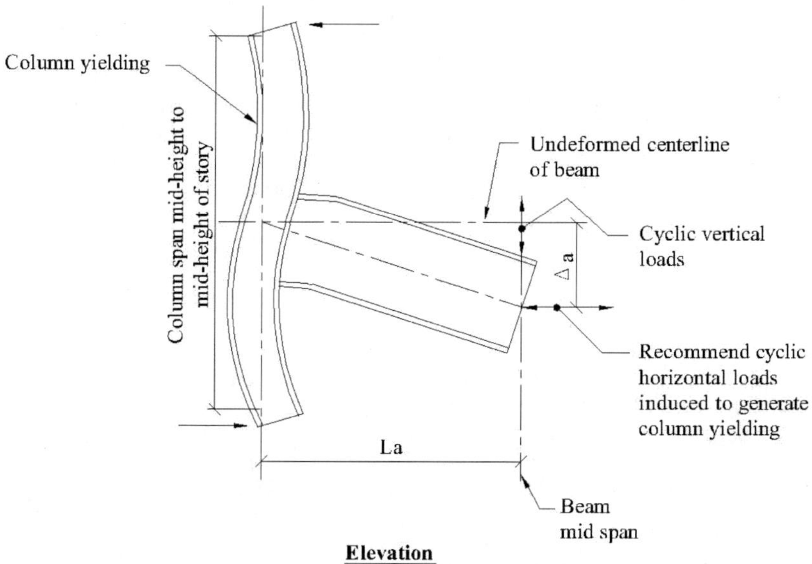

Recommended Test for Steel Moment Frame Connections

Figure 5.1

As mentioned previously in Chapter 1, Section 1.2.11, Partridge et al [Partridge et al (2000)] cited low cycle fatigue as a major cause for the steel moment frame connections in the 1994 Northridge Earthquake and carried out tests demonstrating this. Kanvinde and Deierlein (2005) and Fell et al (2005) have developed procedures to predict fracture due to Ultra Low Cycle Fatigue where extremely large strains with very few cycles occur. Their approach evaluates a Fracture Index at any point in the loading history derived from the state of stress, the plastic strain, the critical void size (see Chapter 2, Section 2.2.3) and a material parameter. Along with finite element computer analysis, including second order effects that simulate local buckling, failure may be accurately predicted. It is interesting to note that Bertero and Popov (1965), over four decades ago, recognized the importance of low cycle fatigue during seismic events. The testing by Bertero and Popov consisted of cyclic load tests on cantilevered 10cm (4 inch) deep wide flange members. Cyclic testing varied from the elastic to the inelastic range. The cyclic testing in the inelastic range caused significant local buckling and torsional displacements, at the clamped end of the beam, that, in the early stages, resulted in cracking that led to the fracture of the beam. Bertero and Popov considered it necessary to prevent local buckling in order for the beam to resist a large number of cycles at the same magnitude of loads. They advised that "in structural work involving repetition of fully or partially reversible loading conditions, the problem of preventing local buckling of elements is considerably more important than which concerns the low fatigue endurance of the material itself".

There has also been testing and development on improved lateral resisting systems such as buckling restrained braced frames (BRBF) which tests show encouraging performance and much improvement with respect to conventional braced frames. However, although the buckling restrained braces have performed well, problems with gusset plates have been found in a few cases. Thus for structures subjected to seismic forces, careful attention to details, to minimize stress and strain levels, appears imperative in development, testing and design particularly where significant stresses yielding is anticipated.

Regarding research and testing on seismic resisting systems, a very comprehensive dissertation and important resource on "Recommendations for Experimental Studies on the Seismic Behaviour of Steel Components and Materials" was published by Krawinkler et al [Krawinkler et al (1983)]. The document considers weldment failures, local buckling, size affects, constant amplitude cycling, variable amplitude cycling, strain rate issues, etc. Krawinkler et al's document written over two decades ago, discusses how low cycle fatigue and plastic fracture mechanics approaches can be used to predict the number of cyclic deformations to failure. Krawinkler et al also give recommendations for component testing for different types of failure modes (e.g. beam and column bending, crack propagation and fracture at beam-to-column moment connections, shear behavior of beam-to-column joints, fracture at net section of bolted moment splices, fracture at net sections of bolted brace connections). According to Krawinkler et al, size effects have a strong influence on failure caused by crack propagation but have less importance with regard to local and lateral torsional buckling modes.

In the opinion of the author that much more comprehensive testing is necessary to establish more clearer recommendations for various loading conditions (low cycle fatigue, high cycle fatigue and blast loading) including welding, bolting and materials taking into account size effects, residual stresses, constraint issues, local buckling, etc.

5.1.6 Environmental conditions

Environmental conditions, their effects, in particular corrosion, are quite well understood. However, aggressive environments have sometimes been more severe than anticipated. Most codes give only brief requirements and engineers may only have rudimentary knowledge of the effects of environmental conditions. Thus experts in these fields should be consulted particularly where aggressive conditions may occur.

5.2 RECOMENDATIONS

5.2.1 Major Issues

Summarizing the major issues discussed in Section 5.1, the author's recommendations may be summarized as follows:

- There is a need to learn from past failures including those from other industries.

- Academic establishments should incorporate at least primer courses in Fracture Mechanics and Metallurgy in their curriculum for Structural Engineering degrees.

- There should be a national research program to further the work carried out to date (e.g. The SAC project) to establish the material, quality control and quality assurance requirements for the different demands on structures (e.g. wind, earthquakes, blast, vibratory loads). The research should include comprehensive testing on low cycle and high cycle fatigue configurations, loading rates, the effects of blasts, variable cyclic loading applied to various joint details (both by welding and bolting), assemblies and varying member thicknesses. Furthermore, it is the opinion of the writer that further research on the effects of corrosion on structures subject to high cycle fatigue, utilizing fracture mechanics considerations, should be encouraged.

5.2.2 Considerations to reduce failures

Design:

- Understand the demands imposed on the structure. Engineers should be encouraged to go beyond mere compliance with Codes. (A summary of typical demands and issues of structures is given in Appendix A).
- Use appropriate computer analysis to provide improved prediction of structural behavior. An example would be use of time history analysis for earthquakes which invariably show significantly higher moments in columns than otherwise predicted by static analysis [Bondy (1996)].
- Limit stress levels (e.g. for high cycle fatigue).
- Limit strain levels where yielding can occur (e.g. for structures subject to seismic events).
- Design details to limit stress and strain concentrations.
- Protect constrained areas from being required to yield.
- Consider loading (strain) rate.
- Avoid welding to areas of steel with low ductility and or fracture toughness (e.g. K areas of wide flange members, corners of hollow steel sections).
- Provide adequate protection from the environment (e.g. provide measures to prevent galvanic action between dissimilar metals.
- Consider the use of devices to reduce demands on the structure. For example, base isolators and/or damping devices for structures in seismic areas; tuned damping devices for structures subject to wind and/or high cycles.
- Consider the secondary stresses associated with residual stresses, eccentricities (both those built in and those arising from fabrication tolerances and weld distortion).

Appendix A gives a listing of demands on steel framed buildings. It should be recognized that the above items represent only a generalized list. Again, as previously stated, engineers are encouraged to research thoroughly the issues which their structures may be subjected to.

Materials:

- Recognize the material is not homogenous and can have significant variation in its properties including yield strengths, ultimate tensile strengths, ductility, fracture toughness, chemistry.
- Recognize grain orientation associated with rolling processes.
- Consider adequacy of material properties for applications (e.g. fracture toughness for low temperature environment).
- Consider the residual stresses due to steel making processes.
- Select materials based upon their limitations (e.g. consider possible low toughness and ductility in wide flange K area and the weld seams and corners of cold formed hollow steel sections).
- Select materials appropriate for the environmental conditions.

Fabrication:

- Utilize details to reduce stress and strain concentrations.
- Use cutting techniques and procedures that do not significantly affect material properties.
- Consider cold forming and heat straightening procedures and their affects on the mechanical properties of steel.
- Utilize grinding at connections where significant stress and strain concentrations may occur (control of surface roughness).
- Use sufficient quality control to ensure accuracy to minimize errors, improper procedures, etc.

Welding:

- Ensure adequate fracture toughness of the weld filler metal to minimize crack propagation.
- Use low hydrogen electrodes.
- Provide welding procedure specifications that have good heat input control.

- Consider procedures to maintain good material properties in the heat affected zone.
- Consider the use of preheat and post heat particularly for thick members.
- Consider weld shrinkage and residual stresses particularly with regard to the sequencing of welding.
- Comply with fit ups and weld profiles.
- Weld details to minimize notch effects and residual stresses.
- Weld details to minimize the potential for lamellar tearing.
- Avoid intermixed welds unless verified by testing.
- Select materials and welding processes that make weld areas cathodic for severe corrosion conditions.
- Provide sufficient inspection, including non-destructive testing to minimize defects, check material for lamellar inclusions, and ensure good weld material properties, etc.

Bolting:

- Design layout to prevent bolt shear failures, including spacing and edge distance to promote yielding of steel material.
- Use proper installation of slip critical bolts.
- Comply with specified bolts, nuts and washer requirements.
- Use proper procedures to ensure plies are tight without gaps.

5.2.3 Relationship of Design, Materials, Quality Control, & Quality Assurance

Figure 5.2 shows the relationship of design, specification of materials, quality control and quality assurance as part of a Fracture Control plan (FCP). By design, engineers can seek to control stress and strain concentrations and protect constrained regions from being required to yield in the case of structures subject to seismic demands. Specification of materials should be such as to ensure adequate ductility, yield levels, material properties including fracture toughness to meet the maximum demands necessary. Sufficient quality control and quality assurance should be provided to ensure material

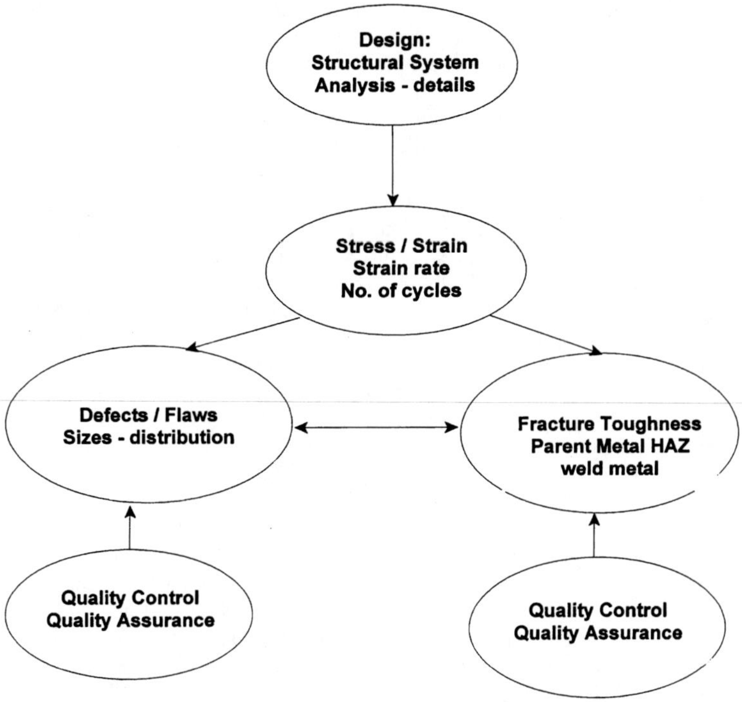

Design, Quality Control and Quality Assurance

Figure 5.2

specifications are met; fabrication including welding procedure specifications and bolt installation is carried out correctly along with non-destructive testing and visual inspection to minimize defects.

This approach is essentially that adopted in FCPs which have been used in industries such as the Shipbuilding Industry. Essentially, a FCP utilizes Fracture Mechanics and testing to establish the specific toughness requirements for the anticipated stress and strain levels, loading rate, joint location and expected flaw size.

The FCP is then implemented by codes and/or construction specifications. An excellent discussion on the use of FCP is given by Williams (1998) regarding Steel, Welding and Construction Specifications for Seismic Structures where he recommends a four tier system for different levels of demands such that

Tier 1 represents the majority of steel fabrication and Tier 4 is for structures subject to load beyond yield strength and/or subject to corrosion, in-service fatigue or other degradation.

Barsom and Rolphe (1999) also give a very comprehensive discussion on Fracture and Fatigue Control. Included in their book is an in-depth discussion on applying Fracture Mechanics Methodology to the incidents of the cracking that occurred in 69 tanker vessels between 1984 and 1988 described in Chapter 1, 1.2.10. Barsom and Rolphe (1999) discuss in depth the methodology which includes identifying the suspect details, analyzing the stress conditions to establish the stress intensity factor and carrying out inspections to establish representative flaw sizes for fatigue crack propagation analysis. They were able to determine representative fracture toughness from maximum stress initial crack size and the number of cycles it takes to grow from the initial crack size to the critical crack size that would cause complete failure of the tanker. Based upon this assessment, reasonable inspection intervals for safe and reliable service were established.

The International Institute of Welding issued a report by a Joint Working Group entitled "IIW Recommendations for assessment of risk of fracture in seismically affected moment connections" [IIW (2003)]. Included in their document is an approach based upon a Risk Assessment Procedure using Fracture Mechanics. They consider qualitative risk assessments procedures based upon CVN toughness in the weld meld, heat affected zone and parental metal. Level I assessment, which is thought to be conservative, indicates low risk for CVN of 135 joules (100 ft.lbs.), medium risk for CVN of 36 joules (27 ft.lbs.) and very high risk for CVN of 14 joules (10 ft.lbs.). Level II assessment, which is more in-depth and is based upon fracture mechanics methods, evaluates the minimum service temperature or fracture toughness requirements for high toughness, medium toughness and low toughness requirements.

5.3 CLOSING COMMENTS

As previously mentioned, the intent of this document has been to help provide a general understanding of the issues associated with brittle and fatigue failures, and issues associated with corrosion. Furthermore, it is intended to provide guidance, promote interest and provide recommended references for further study and elucidation on these subjects. With this document, an opportunity was available to express concerns regarding current practice and give recommendations.

APPENDIX A

DEMANDS ON STEEL FRAME BUILDINGS

Demands on Steel Frame Buildings can vary from static loads with stresses essentially within the elastic range, to significant yielding such as may occur in a seismic event or during a blast. The load/force application may be cyclic in nature and strain rates could vary from slow to rapid.

Demands on steel structures and issues associated can be briefly summarized as follows:

A1.1 **High Seismic Areas**:

A seismic event can cause significant yielding to occur during a number of cycles.

Moment Frames

Demands

- Stress/Strain Concentrations
- Determining the affects of higher modes (i.e. column moment magnification) [Bondy (1996) found from time history analysis substantially higher moments in the columns than predicted by conventional static analysis.]
- Low Cyclic Behavior
- Post Yield Stress
- Very High Strains
- Intermediate Strain Rate

Issues

- Properties of Welds
    - Strength (overmatching/undermatching)
    - Significant Ductility
    - Toughness
- Properties of Parent Metal
    - Strength
    - Significant Ductility
    - Toughness
- Stress and Strain Concentrations (e.g. at weld interface)
- Triaxial Constraint (e.g. at through thickness conditions)

- Weld Residual Stresses Due to Constraints
- Parent Metal Residual Stresses
- Access Hole Configurations
- Bolt Slippage
- Panel Zone Yielding
- Local Buckling of Flanges and Web
- Lateral Torsional Buckling
- Lateral Bracing, Design Forces/Stiffness

Braced Frames

Demands
- Post Yield Stress
- Moderately High Strains
- Low Cyclic Behavior
- Intermediate Strain Rate

Issues
- Properties of Welds
  - Strength (overmatching/undermatching)
  - Significant Ductility
  - Toughness
- Properties of Parent Metal
  - Strength
  - Ductility
  - Toughness
- Low Cycle Fatigue
- Stress and Strain Concentrations (e.g. at connections)
- Triaxial Constraint
- Weld Residual Stresses Due to Constraints
- Parent Metal Residual Stresses
- Access Hole Configurations
- Bolt Slippage
- Local Buckling of Flanges and Web
- Post Buckling Behavior
- Beam Stiffness at braced frames and their effects on strain rate

A1.2 High Winds & Repetitive Winds:

High winds may cause high stresses although stresses are unlikely to exceed yield. The duration of the high winds can be lengthy and thus can result in numerous cycles. The possibility of brittle failure or low cycle fatigue exists.

Moment Frames and Braced Frames

Demands
- Usually Below Yield Stress in High Winds
- Low Cyclic Behavior in High Winds
- High Cyclic Behavior at Low Stresses in Repetitive Winds

- Low Strain Rate

Issues

- Properties of Welds
  - Strength (overmatching/undermatching)
  - Ductility
  - Toughness
- Properties of Parent Metal
  - Strength
  - Ductility
  - Toughness
- Low Cycle Fatigue (high winds)
- High Cycle Fatigue (repetitive)
- Stress and Strain Concentrations
- Triaxial Constraint

Secondary Stresses

- Weld Residual Stresses Due to Constraints
- Parent Metal Residual Stresses

A1.3 **Catastrophic Event Leading to Progressive Collapse:**

Demands

- Post Yield Stress
- Very High Strains
- Dynamic Behavior
- Very High Strain Rate (effecting material behavior)

Issues

- Properties of Welds
  - Strength (increases significantly due to high strain rate) [However, there is an appreciable shift (increase) in the nil ductility temperature.]
  - Ductility
  - High Toughness
- Properties of Parent Metal
  - Strength (increases significantly due to high strain rate) [However, there is an appreciable shift (increase) in the nil ductility temperature.]
  - Ductility
  - High Toughness
- Stress and Strain Concentrations
- Triaxial Constraint

Secondary Stresses

- Access Hole Configurations
- Local Buckling of Flanges and Web
- Lateral Torsional Buckling
- Lateral Bracing

References

Adams, C.J. (1975). "The Short Transverse Fatigue Properties of Steel", Division of Structural Engineering and Structural Mechanics, University of California, Berkeley, June 11.

(AISC) (1930). "Steel Construction", 1st Edition. American Institute of Steel Construction, April

AISC (1950). "Steel Construction", Fifth Edition, 1950. American Institute of Steel Construction

AISC (1973). "Commentary on Highly Restrained Connections," American Institute of Steel Construction, Journal, Third Quarter,

AISC (2005). "Steel Construction Manual, 13th Edition." American Institute of Steel Construction

AISC (2007). "AISC Requests Data from Independent Testing", Modern Steel Construction, American Institute of Steel Construction, November edition.

Anderson, J.C., Duan, J., Xiao, Y. and Maranian, P., (2000). "Improvement of Welded Connections Using Fracture Tough Overlays". Department of Civil and Environmental Engineering, University of Southern California, CA, USA.

AIJ (1995). "Preliminary Reconnaissance Report of the 1995 Hyogoken-Nanbu Earthquake." English Edition, April. Architectural Institute of Japan, 5-16-20, Shiba, Minato-ku, Tokyo, Japan.

Avent, R.R. and Mukai, D.J., (2001). "What You Should Know About Heat Straightening Repair of Damaged Steel." American Institute of Steel Construction, Engineering Journal, First Quarter, 2001.

ASTM A500 (2001). "Standard Specification for Cold-Formed Welded and Seamless Carbon Steel Structural Tubing in Rounds and Shapes." A500-01, American Society for Testing and Materials, West Conshohocken, PA, U.S.A.

AWS D1.1 (2004). "Structural Welding Code – Steel." The American Welding Society.

AWS D1.1 (2006). "Structural Welding Code-Steel." The American Welding Society.

Banovic, S.W. and Siewert, T.A. (2006). "Characterization of Submerged Arc Welds from the World Trade Center Towers As-Deposited Welds and Failures Associated with Impact Damage of the Exterior Columns". The American Welding Society, The Welding Journal, July.

Barsom, J.M. and Korvink, S., (1997). "Through Thickness Properties of Structural Steel." Report No. lSAC/BD-97/01, July 1, SAC Joint Venture.

Barsom, J.M. and Rolphe, S.T, (1999). "Fracture and Fatigue Control in Structures." 3rd Edition, American Society for Testing and Materials, ISBN O-8031-2082-6.

Bartlett, F.M., Dexter, R.J., Graeser, M.D., Jelinek, J.J., Schmidt, B.J., Galambos, T., (2003). "Updating Standard Shape Material Properties Database for Design and Reliability." American Institute of Steel Construction, Engineering Journal, First Quarter.

Beedle, L.S. and Tall, L., (1959). "Basic Column Strength", Proceedings of the American Society of Civil Engineers, July.

Bennett, P.E., Sinclair, G.M., (1965). "Parameter Representation of Low-Temperature Yield Behavior of Body-Centered Cubic Transition Metals." ASTM paper 65-MET-11.

Bertero, V.V., Anderson, J. C. and Krawinkler, H. (1994). "Performance of Steel Building Structures During the Northridge Earthquake." College of Engineering, University of California Berkeley.

Bertero, V.V. and Popov, E.P. (1965). "Effect of Large Alternating Strains of Steel Beams." Journal of the Structural Division, Proceedings of the American Society of Civil Engineers, February.

Bjorhoude, R., (2006). "Cold Bending of Wide Flange Shapes for Construction." American Institute of Steel Construction, Engineering Journal, Fourth Quarter.

Blodgett, O.W., (1998). "The Effects of Constraints on Ductility in Welded Beam to Column Connections." International Conference on Welded Construction in Seismic Areas, Maui, Hawaii; The American Welding Society, October.

Bondy, K.D., (1996). "A More Rational Approach to Capacity Design of Seismic Moment Frame Columns." Earthquake Spectra, Earthquake Engineering Research Institute, Oakland, California, August.

Bonowitz, D., Youssef, N,. (1995). "Steel Moment-Resisting Frames After Northridge." Modern Steel Construction, American Institute of Steel Construction, May.

Boyd, G.M., (1970). "Brittle Fracture in Steel Structures." Butterworth, London.

British Standard 7910. "Guidance on Methods for Assessing the Acceptability of Flaws in Fusion Welded Structures." The British Standards Institute.

Bruneau, M., Bisson, M., Sarrat, M., (1998). "Tests of Three Seismic Retrofit Schemes for Riveted Stiffened Seat Angle Connections." 6th U.S. National Conference on Earthquake Engineering.

Brandow, G.E. and Maranian, P., (2001). " Methods of Repair of Damaged Existing Welded Steel Moment Frame Buildings Including Weld Overlay Repair Techniques." Structural Faults and Repair, 9th International Conference, London, United Kingdom (available from the University of Edinburgh, Scotland).

Burdekin, M., (1999). Gold Metal Lecture. "Why Size Matters in Large Structures." The Structural Engineer, The Institution of Structural Engineers. Vol. 77/No. 20, October (United Kingdom).

Burdekin, M. and Suman, A,. (1998). "Further Thoughts on the Relationship Between Toughness, Workmanship and Design for Earthquake Resistant Structures." University of Manchester Institute of Science and Technology, United Kingdom.

Byfield, M.P. and Nethercot, D.A., (1997). "Material and Geometric Properties of Structural Steel for Use in Design." The Structural Engineer, The Institution of Structural Engineers, 4 November (United Kingdom).

Campus, F,. (1954). "Effects of Residual Stresses on the Behavior of Structures in Metals and Metal Construction." W.R. Osgood, Ed. Reinhold Publishing Co, 1-21.

Civil Engineering (Feb. 2008). "Underdesigned Gusset Plates Cited by NTSB in I-35W Bridge Collapse Investigation" by Laurie A. Shuster, The Magazine of the American Society of Civil Engineers, Volume 78, No. 2.

Darcis, Ph., P., McCloskey, J.D. McCowan and Siewart, T.A., (2007). "Exploring Methods for Measuring Pipe Weld Toughness." The American Welding Society, Welding Journal, June.

Division of the State (California) Architect (DSA), (2007). "Defective Product Alert." Structural Tube Steel, DSA Bulletin 07-03 Update, Available from DSA website, October 29.

Dong, P. and Zhang, J. (1998), "Residual Stresses in Welded Moment Frames and Implications on Structural Performance."International Conference on Welded Construction in Seismic Areas, Maui, Hawaii; American Welding Society, October.

Dowling, N. E., (1999). "Mechanical Behavior of Materials-Engineering Methods for Deformation, Fracture and Fatigue." 2nd ed., Prentice Hall.

Dowling, N.E. (2007); "Mechanical Behavior of Materials – Engineering Methods for Deformation, Fracture and Fatigue", 3rd edition, Prentice Hall.

Engelhardt, M.O., Winneberger, T., Zekany, A.J., and Potraj, T.J., (1996). "The Dog Bone Connection, Part II", Modern Steel Construction, American Institute of Steel Construction, August.

European Standard En 10164. "Steel Properties With Improved Deformation Properties Perpendicular to the Surface of the Product." ISBN: 0580451186.

Farrar, et al, (1975). "Investigations into lamellar tearing." A compendium reports from a sponsored research programme Phase I-IV and Phase V-VII; The Welding Institute, England.

Farrar, J.C. and Dolby. R.E, (1972). "Lamellar Tearing in Welded Steel Fabrication." The Welding Institute, England.

Fell, B.V., Myers, A.T., Deierlein, G.G. and Kanvinde, A.M. (2005). "Testing and Simulation of Ultra-Low Cycle Fatigue in Steel Braces." 8th National Conference on Earthquake Engineering, San Francisco, April.

FEMA 267, (1995). "Interim Guidelines: Evaluation, Repair, Modification and Design of Welded Steel Moment Frame Structures." Federal Emergency Management Agency, August.

FEMA 267A, (1997). "Interim Guidelines Advisory No. 1, Supplement to FEMA 267, Federal Emergency Management Agency, January.

FEMA 274, (1997). "NEHRP Commentary on the Guidelines for the Seismic Rehabilitation of Buildings." Federal Emergency Management Agency, October.

FEMA 288, (1996). "Background Reports: Metallurgy, Fracture Mechanics, Moment Connections and Frame Systems Behavior", Federal Emergency Management Agency.

FEMA 350, (2000). "Recommended Seismic Design Criteria for New Steel Moment-Frame Buildings." Federal Emergency Management Agency, June (Also, FEMA 350 Errata, March 16, 2001)

FEMA 351, (2000). "Recommended Seismic Evaluation and Upgrade Criteria for Existing Welded Steel Moment-Frame Buildings." Federal Emergency Management Agency, June.

FEMA 352, (2000). "Recommended Post-Earthquake Evaluation and Repair Criteria for Welded Steel Moment-Frame Buildings," Federal Emergency Management Agency, June.

FEMA 353, (2000). "Recommended Specifications and Quality Assurance Guidelines for Steel Moment Frame Construction for Seismic Applications," Federal Emergency Management Agency, June.

FEMA 355D, (2000). "State of the Art Report on Connection Performance," Federal Emergency Management Agency, September.

Fisher, J.W., Kulak, G.L., Smith, I.F.C., (1998). "A Fatigue Primer for Structural Engineers." National Steel Bridge Alliance, 102-106.

Fisher, J.W., Kulak, G.L., Smith, I.F.C., (1997). "A Fatigue Primer for Structural Engineers." Advanced Technology for Large Structural Systems (ATSS) Report No. 97-11, October.

Funderburk R. Scott, (2000). "Taking your Weld's Temperature." Modern Steel Construction. American Institute of Steel Construction, February.

Galambos, T.V. and Ravindra, M.K., (1978). "Properties of Steel for use in LRFD." Journal of Structural Engineering, The American Society of Civil Engineers, September.

Gale Research Inc. (1994). "When Technology Fails, Significant Technological Disasters Accidents and Failures of the Twentieth Century." Edited by Neil Schlager.

Gartzke, W. H., Yoerger, D. R., Harris S., Dublin, R. O. and Brown, D. K. (1992). "Deep Underwater Exploration Vehicles - Past Present and Future." Paper Presented at the Centennial Meeting of the Society of Naval Architects and Marine Engineer, New York City.

Grover, L.M. (1954). "Design and Research for Welded Structures." The American Society of Civil Engineers, Volume 79, Separate No. 343, November.

Hamburger, R. and Meyer, J., (2006). "The Performance of Steel-Frame Buildings with Infill Masonry Walls in the 1906 San Francisco Earthquake." Earthquake Spectra, Special Issue II, Vol. 22, Earthquake Engineering Research Institute. April.

Hansen, B. (2007)." Forensic Engineering, Bridge Collapse Prompts Investigations, Raises Questions; Civil Engineering, Vol. 77, No. 9." The American Society of Civil Engineers, September.

Harrison, P.L. and Webster, S.E., (1995). "Examination of Two Moment Resisting Frame Connectors Utilizing a Cover Plate Design." British Steel Technical.

Holt, R. and Hartmann, J. (2008). "Adequacy of the U10 & L11 Gusset Plate Designs for the Minnesota Bridge No. 9340 (I-35W over the Mississippi River)." Federal Highway Administration, Turner-Fairbank Highway Research Center Report.

"International Institute of Welding (IIW)(2003), JWG of Commissions X and XV-G (IIW) (2003)". "Recommendations for Assessment of Risk of Fracture in Seismically Affected Moment Connections." IIW Doc XV-1102-03, Prepared by the Joint Working Group of Commissions. X and XV-G of the International Institute of Welding Published in Welding in the World, Vol. 47, no. ¾, March.

Ishikawa, N., Kobayashi, Y. and Kurihara, M. and Toyoda, M., (1998). "Ductile-Brittle Transition Behavior of Structural Steels Differing in Ductility Under Dynamic Loading." International Conference on Welded Construction in Seismic Areas; Maui, Hawaii; The American Welding Society, October.

Kaminetzky, D., (1991). "Design and Construction Failures, Lessons from Forensic Investigations." McGraw-Hill, Inc.

Kanvinde, A.M. and Deierlein, G.G. (2005). "Continuum Based Micro-Models for Ultra Low Cycle Fatigue Crack Initiation in Steel Structures". Proceedings, The American Society of Civil Engineers Structures Congress and Exposition, New York, NY, 2005.

Krawinkler, H., Zohrei, M., Lashkarin, I.B., Cofie, N.G., Hadidi-Tamjed, H., (1983). "Recommendation for Experimental Studies on the Seismic Behavior of Steel Components and Materials". The John A. Blume Earthquake Engineering Center, Department of Civil Engineering, Stanford University, Stanford, California 94305. September.

Kosteski, N., Packer, J.A. and Puthli, R.S. (2005). "Notch, Toughness of Internationally Produced Hollow Structural Sections." Journal of Structural Engineering ASCE, February.

Lancaster, J., (1992). "Handbook of Structural Welding." McGraw Hill.

Lancaster, J., (1996). "Engineering Catastrophes, Causes and Effects of Major Accidents." John Lancaster; Abington Publishing (United Kingdom).

Lancaster, J., (1997). "Handbook of Structural Welding." Woodhead Publishing Ltd., (United Kingdom).

Lazor, R., Dinovitzer, A, Begg, D., Pussegoda, N. and Semiga, V., (2005). "Delayed Cracking in Multipass Welds." The American Welding Society, Welding Journal September.

Levy, M. and Salvadori, M., (1992). "Why Buildings Fall Down." ISBN 0-393-03356-2, W.W. Norton & Company.

Lindley, C., Bateson, P.H., Bannister, A.C. and Pike, T.J., (2001). "Contributions on the Lamellar Tearing and Fracture Resistance of Heavy Sections for use in Seismic Loading Conditions." International Institute of Welding Doc-XV-G-100-2001.

Maranian, P., (1997). "Vulnerability of Existing Steel Framed Buildings Following the 1994 Northridge (California USA) Earthquake." Considerations for Their Repair and Strengthening; The Structural Engineer, The Institution of Structural Engineers, May 20. (United Kingdom)

Maranian, P. and Simon, W., (2002). "Static Small Component Tests on 2-inch-thick Specimens Using Weld Overlays." The American Society of Civil Engineers, Journal of Materials in Civil Engineering, Vol. 14, No. 1, January/February.

Masubuchi, K., (1980). "Analysis of Welded Structures." Massachusetts Institute of Technology, Pergamon Press.

McCarthy, J.H. and Foecke, T., (2007); What Really Sank the Titanic, C.S.I.

McEvily, A.J., (2001). "Metal Failures, Mechanisms Analysis, Prevention." Wiley-Interscience.

Miller, D.K. (1993). "The Challenge of Welding Jumbo Shapes Part 1." The American Institute of Steel Construction Specifications; The Welding Innovation Quarterly, Volume X, No. 1.

Munse, W.H., (1964). "Fatigue of Welded Steel Structures." Welding Research Council, 46 (Statement on Size Effect).

New Civil Engineer (NCE) International (2007): Magazine of the Institution of Civil Engineers, "URS in I-35 Legal Threat", August..

New Yorker (1995). "The Fifty Nine Story Crisis." Published in May 29, 1995

Norwegian Institute of Technology. "The Progressive Failure of the Alexander L. Kielland Platform CISM." Courses and Lecture No. 283. International Centre for Mechanical Sciences.

Pargeter R. (2003). "Evaluation of Necessary Delay Before Inspection of Hydrogen Cracks." The American Welding Society, Welding Journal, November.

Parker E.R. (1957). "Brittle Behavior of Engineering Structures." Published by John Wiley & Sons Inc., 254 (Quotation from Engineering News).

Partridge J.E., Paterson S.R., Richard R.M., 2000. "Low Cycle Fatigue Tests and Fracture Analyses of Bolted-Welded Seismic Moment Frame Connections." July; STESSA 2000, Third International Conference, Montreal, Canada.

Patel, P. (2006). Technical Notes, "Welding Preheat" Published by The Smith Emery Company, Los Angeles, California.

Patel, P. (2008). "Technical Notes, Analysis of Failures". Published by the Smith Emery Company, Los Angeles, California.

Pellini (1971). "Principles of Fracture Safe Design Part I." Supplement to the Welding Journal, Welding Research Council, March.

Pellini (1971). "Principles of Fracture Safe Design Part II." Supplement to the Welding Journal, Welding Research Council, April.

Perdomo, J.J., Spry, T.D., Indacochea, J.E., Polar, A. and Rumiche, F. (2005). "Weld Cracking Linked to Wires Containing Boron." The American Welding Society, Welding Journal, November.

Pribula, M. (2006). "Fracture Critical Bridge Inspection, Bridge #9340, I-35W over the Mississippi River at Minneapolis, June." Minnesota Department of Transportation Metro District Maintenance Operations, Bridge Inspection, 10 ("Fracture Critical")

Quintanna, M.A. and Johnson, M.Q. (1997). "The Effects of Intermixed Weld Metal on Mechanical Properties – Part 1", presented at The American Welding Society Annual Convention.

Quintanna, M.A. and Johnson, M.Q. (1998). "The Effects of Intermixed Weld Metal on Mechanical Properties – Part 2", presented at The American Welding Society Annual Convention.

Quintanna M.A. and Johnson M.Q. (1998). "The Effects of Intermixed Weld Metal on Mechanical Properties – Part 3." International Conference on Welded Construction in Seismic Areas, Maui, Hawaii; The American Welding Society, October.

RCSC (2001). Kulak G.L., Fisher J.W., Struik J.H.A. "Guide to Design Criteria for Bolted and Riveted Joints." The Research Council on Structural Connections; American Institute of Steel Construction.

Richard, R.M.; Partridge, J.E., Allen, J. and Radua, S. (1995). "Finite Elements Analysis and Tests of Beam to Column Connections." Modern Steel Construction October.

Ricles, J.M., Mao, C., Lu, L.W., and Fisher, J. (2000). "Development and Evaluation of Improved Details for Ductile Welded Unreinforced Flange Connections." SAC/BD-00/24. SAC Joint Venture.

Rollason, E.C. (1973). "Metallurgy for Engineers." 4th Edition Published by Edward Arnold.

Sarkkinen, D.L., (1998). "Through Thickness Tensile Stress in Hot-Rolled Steel Members." Elsevier Science Ltd.

SEAOC (2002). "Commentary and Recommendations on FEMA 350" by the Structural Engineers Association of California, Seismology Committee, FEMA 350 Task Group (available from the SEAOC web site).

SEAOSC (2005). Presentation by Dr. Patxi Uriz at the Structural Engineers Association of Southern California Steel Macro Seminar on March 20 (available from SEAOSC).

Shimanuki H. and Hagiwara Y. (1998). "Fracture Mechanics Analysis of Damaged Steel-Framed Structures in Recent Earthquakes." International Conference on Welded Construction in Seismic Areas, Maui, Hawaii; The American Welding Society, October.

Sih C.G. (1989). "Case Study of Point Pleasant Bridge Collapse: Cause of Failure." International Conference on Case Histories in Structural Failures; Singapore, March

Simon W., Anderson J., Compton J., Hayes W. and Maranian P., (1999). "Repair of Existing Steel Moment Frame Buildings Damaged From Earthquakes Using Fracture Tough Weld Overlays." American Institute Engineering Journal, Fourth Quarter.

Simon W., Anderson J., Compton J., Hayes W., Maranian P., (1998). "Dynamic Load Welds for Repair of Existing Steel Moment Frame Buildings Damaged from Earthquakes." DLW Rep. No. 1341 (Available from Brandow & Johnston, Inc., 444 South Flower Street, Los Angeles, California, 90071, (U.S.A.).

SSPC (1994). Statistical Analysis of Tensile Data for Wide Flange Structural Shapes. Structural Steel Producers Council.

Steel Designer's Manual, 4th Edition, (1972) Construction Steel Research and Development Organization Pub. Crosby Lockwood Staples, London, England

Stout, R.D., (1987). "Weldability of Steels." 4th Edition, Welding Research Council.

Tobriner, S. (2006). "An EERI Reconnaissance Report: Damage to San Francisco in the 1906 Earthquake – A Centennial Perspective." Earthquake Spectra, Special Issue II, Vol. 22, Earthquake Engineering Research Institute, Oakland, California, April.

Torroja, E. (1958). "Philosophy of Structures." University of California Press, Berkeley and Los Angeles, California, 48 & 49.

Tsai C.L., Han M.S. and Jung G.H. (2006). "Investigating the Bifurcation Phenomenon in Plate Welding." The American Welding Society. The Welding Journal, July.

Tsai, C.; Kim, D.; Jaeger, J.; Shim, Y.; Feng, Z.; and Papritan, J. (2001). "Design Analysis for Welding of Heavy W Shapes." The American Welding Society, The Welding Journal, February.

Uriz, P., and Mahin, S., (2004). "Seismic Performance Assessment of Concentrically Braced Steel Frames." Proceedings of the 13th World Conference on Earthquake Engineering.

Williams [1998]. "Steel, Welding and Construction Specifications for Seismic Structures." International Conference on Welded Construction in Seismic Areas, Maui, Hawaii. The American Welding Society, October.

Withey, M.O., (1928). "Tests of Specimens Cut From Different Portions of Structural Steel Shapes." Proceedings of ASTM., Vol. 28, Part II.

Yee, R.K., Egan, G. and Paterson, S.R. (1998). "Engineering Evaluations of Column Continuity Plate Detail Design and Welding Issues in Welded Steel Moment Frame Connections." International Conference on Welded Construction in Seismic Areas, Maui, Hawaii, The American Welding Society, October.

Background Reading Material

AISC Seismic (2005). "Seismic Design Provisions for Structural Steel Buildings." American Institute of Steel Construction.

Althouse, A.D., Turnquist, C.H., Bowditch, W.A. Bowditch, K.E. (2000). "Modern Welding." The Goodheart-Willcox Company Inc.

ASTM A325 or A490 (2000). "Specification for Structural Joints Using ASTM A325 or A490 Bolts, Research Council on Structural Connections." American Institute of Steel Construction, Inc. June 23.

AWS D1.8 (2006). "Structural Welding Code – Seismic Supplement." The American Welding Society.

Blodgett, O.W., Funderburk, R.S., and Miller, D.K. (1999). "Fabricator's and Erector's Guide to Welded Steel Construction." James F. Lincoln Arc Welding Foundation.

Collins, J.L., (2008). "The Hole Story." Modern Steel Construction, The American Institute of Steel Construction, June.

Davies, B.J. and Crawley, E.J. "Structural Steelwork Fabrication." Volume One, The British Constructional Association Ltd., Bournehall Press Ltd.

FEMA 274, (1997). "NEHRP Commentary on the Guidelines for the Seismic Rehabilitation of Buildings." Federal Emergency Management Agency, October.

FEMA 288, (1996). "Background Reports: Metallurgy, Fracture Mechanics, Moment Connections and Frame Systems Behavior", Federal Emergency Management Agency.

FEMA 355A, (2000). "State of the Art Report on Base Metals and Fracture," Federal Emergency Management Agency, September.

FEMA 355B, (2000). "State of the Art Report on Welding and Inspection," Federal Emergency Management Agency, September.

FEMA 355F. "State of the Art Report on Performance Prediction and Evaluation of Steel Moment-Frame Buildings." Federal Emergency Management Agency, September.

Jones, D.A., (1996). "Principals and Prevention of Corrosion," 2nd Edition, Pub. Prentice Hall, Upper Saddle River, NJ 07458.

Modern Welding Technology, Howard B. Cary, pub. Prentice Hall. 6th Edition, 2004
Modern Welding by Althouse, A.D., Turnquist. C.H., Bowditch W.A, Bowditch K.E, Pub. The Goodheart-Wilcox Company Inc.

Patel, P. (2007). "Welding Preheat." Structural Engineers Association of Southern California (SEAOSC), Steel Macro Seminar, June 9 (available from SEAOSC).

Pratt, J.L. (1979). "Introduction to the Welding of Structural Steelwork." published by Constrado.

Steel Designer's Manual, 5th Edition, (2000) The Steel Construction Institute, Blackwell Science.

Index